THE BACKYARD HOMESTEADING FOR BEGINNERS GUIDE

CREATING A RESILIENT HOMESTEAD OASIS IN ANY SIZE YARD FOR SELF-SUFFICIENT LIVING; SYSTEM DESIGNS WITH GARDENING, GROWING, AND PRESERVING FOOD TECHNIQUES

OWEN ROOTSFIELD

SPECIAL BONUS !

Want This Bonus Book For *FREE* ?

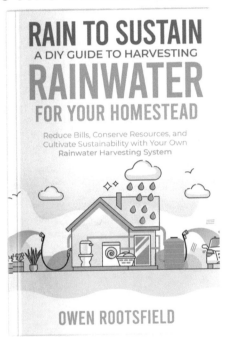

Get **FREE**, unlimited access to it and all of our new books by joining the fan base !

Scan The QR Code With Your Camera

CONTENTS

Part III
RUNNING YOUR HOMESTEAD

INTRODUCTION

Since the world's recovery from Covid-19, people's eyes have opened to how fragile everything is around us. With the lockdown experienced through most parts of the planet, there was a breakdown in getting all the essential items people needed. Yet, Covid-19 isn't the only disaster that can be blamed for what happened.

Poor farming practices have resulted in soil degradation, leading to fewer yields yearly. Climate is also playing a role, with changing weather patterns making it more difficult for farmers to produce the food that is required to keep over seven billion people from starving. The human species has spent too many years being dependent on everyone else around them for their survival.

With the cost of necessities increasing yearly, it's getting more difficult to get by on what is earned. While it may seem like endless doom and gloom, there is a way around the loss of food security and feeling stressed about what is happening in the world today. Have you ever wondered if there was a way to make your life more self-sufficient, to be more dependent on yourself, and to be free from social restraints?

Living a self-sufficient lifestyle is one of the ways you can make a change for the better in your own life. Many people see the act of living self-sufficiently as a way to escape the uncertainty of the future. By living this way, they can guarantee their family will always have food and live a more comfortable life. However, being self-sufficient is more than just taking control of where your food comes from. It's the act of being able to rely on yourself when the times get tough, with little to no outside help, but acknowledging that you can still cooperate with similar-minded individuals.

While many people would love to cut themselves off from the outside world, becoming 100% self-sufficient just isn't possible, as not many people know how to make the necessary tools to survive. Creating a 100% self-sufficient homestead takes years of dedication and self-sacrifice. However, starting an urban or backyard homestead can be just as rewarding and far easier to achieve.

Starting with a backyard homestead comes with many benefits including but not limited to:

- Creating a sustainable food source and lowering your carbon footprint.
- Encouraging polyculture, as growing a variety of plants benefits the environment.
- Helping to change your mindset about waste and recycling.
- Boosting the sense of wellness and your health.

Yet, starting a homestead, whether in your backyard or on a larger piece of property, isn't something that can be done overnight. While there are many success stories on the internet and YouTube (such as Homegrown Handgathered), it's important to remember that this will take time, especially when you need to determine your goals and prepare yourself, both mentally and physically. That is where this book comes in.

Too many people decide to start homesteading, only to start too big and fail soon after. It has taken me decades to accumulate the research through blood, sweat, and tears, to bring to you *The Backyard Homesteading for Beginners Guide: Creating a Resilient Homestead Oasis in any Size Yard for Self-Sufficient Living; System Designs with Gardening, Growing, and Preserving Food Techniques.* Why struggle through developing your homestead from scratch when I have done all the work already?

Dive into these pages to learn more about how to:

- produce ample food, plants, and animals alike
- stay within the letter of the law
- develop the correct mindset to be a homesteader
- understand the work involved and lay the correct foundation
- preserve your own food
- even turn a small profit

Once you have completed this book, not only will you be able to start your homesteading adventure, but you'll be able to start it on a solid foundation, giving you everything you need to build a successful self-sufficient life. While there is some information you may think you know, take the time to dive between each page and start with the basics and avoid potential disappointment and failure. So, turn to the first chapter and dip your toe into the basics of getting a back-yard homestead up and running.

PART I

PLANNING AND DESIGNING YOUR BACKYARD HOMESTEAD

In this section of the book, we'll cover the basics of getting you started on building the idea of a homestead. The following chapters will help you understand what homesteading is and how to design a homestead, and how you can begin to assess all your needs.

FIRST THINGS FIRST

Without a solid foundation, you're going nowhere with your plans to be a homesteader. Before you can get started, it's important to understand the hype surrounding self-sufficiency.

Self-sufficiency is defined as requiring no outside help to fulfill your most basic needs. As it stands, few people have even a single skill to help them be self-sufficient. When self-sufficient, you rely on your physical and mental strength, as well as an array of skills to provide you with a comfortable life. Many believe that living a self-sufficient life is a simpler life with less unnecessary hustle and bustle of the modern lifestyle. While it may be easier to order what you need from an app on your phone, having to work and travel countless hours to earn the money to afford this luxury may be too much for some people.

While many picture it as a life in the middle of nowhere, this is but one type of self-sufficient living. One that doesn't suit everyone's tastes. People turn to a self-sufficient lifestyle to break away from the constraints of modern society, the hustle culture, and the notoriously poor diet, and they become more self-independent and free.

Yet, is homesteading and self-sufficiency easy? Not at all, and if you think it is, then you're already facing a problem. While doomsday preppers love to throw around scary words that make people feel they should be prepared to be 100% self-sufficient, this isn't realistically possible. Even those that live a self-sufficient lifestyle in the middle of Alaska still rely on modern tools and trade to survive. While some are more self-sufficient than others, there is no need to feel driven to completely cut contact with the world and walk into the unknown wilds. This is a sure way to not only fail, but probably die.

Self-sufficiency is about upholding your individuality, independence, and self-determination through learning skills to help you rely less on grocery stores, your doctors (within reason), and even the grid. However, that doesn't mean you shouldn't have a safety net to help you if you get into a difficult spot. Money may not be able to buy everything, but it does help when you can't trade or barter for something you cannot make or grow. You'll need to find a balance between what you're willing to still buy versus what you can produce from your homestead.

WHAT HOMESTEADING REALLY IS

Homesteading means different things to different people. For some, it may mean getting off the grid, while others just want to be able to control where their food comes from and what's in it. For most, it's about living as self-sufficiency and sustainably as possible. This can mean learning to forage, hunt, make your own garden, or raise backyard-friendly livestock. With urban homesteading, it's a good idea to start small and use your garden to grow an array of delicious treats, and if allowed by county or state, get a few backyard-friendly animals.

In the past, the Homesteading Act of 1862 was used as a way to encourage people to travel to the western parts of America. These pioneers were given land on which they could build homes and ground to farm. Some had tremendous success, while others didn't. This act was ended in 1935, as the government was concerned that all the land would be used, and they wanted some of the country set aside for the conservation of nature. Despite this, in the 1970s there was an increase in people interested in homesteading once more, as they had become disillusioned with living in urban or suburban areas. They wanted a better life.

Thankfully, with modern homesteading, there is no longer a need to uproot your family and travel across the country to experience a new life, unless you want to. Nowadays, people can choose to live more sustainably by growing their food, leaving the grid (according to country and state laws), and even treating their illnesses with medicinal herbs instead of running to the doctors for minor inconveniences. Instead of moving away from society, people are moving away from certain aspects of society, be it the education system (home-schooling), medicine (depending on illnesses), the grid (using alternative energy sources), and much more.

IS HOMESTEADING YOUR THING?

Homesteading isn't for everyone. While it doesn't take someone special to become a homesteader, it does take a certain type of person to see the work to the end. While homesteading is a simpler life, it's hardly easy. The more self-sufficient you want to be, the harder you need to work. No one else is going to milk the goats or collect the eggs if you don't do it. However, if you and your family are all willing to make the change to homesteading, it can be a wonderful bonding experience. Plus, it gives you some extra helping hands you'll need.

The type of person who is likely to become a homesteader is someone who is:

- not afraid to work hard
- wants self-sufficiency
- is resilient and resourceful

Those who homesteading is least suited to are those who:

- refuse to be adaptable
- can't be bothered with learning new skills
- are completely dependent on others

Living this way is a steep learning curve, and it's far better to take the time to prepare yourself before attempting it. You need to be in control of every aspect of your life, including debt and addiction. You also need to be the kind of person who can live frugally, while saving money, and be able to live on a shoestring budget if needed. You'll need to be able to make sacrifices, especially on things that will require too much of your energy and time.

However, it isn't all bad. As you start your journey, by learning from this book or other resources, you may feel compelled to compare your successes with the successes of others. You may find that others adapt better than you, while you're struggling. Stop comparing yourself to others. Everyone takes this journey at their own pace, and there's no need for you to start a big project if you're happy to just grow seasonal vegetables in your backyard garden.

You won't be alone on this journey. So, even if you think that you aren't ready for the plunge, take the time to connect with other like-minded individuals. This is how you can learn about the homesteading community and how they operate in terms of making items, producing crops, and even offering services. You may even learn some vital tips from those who have been living this way for decades.

BENEFITS OF BACKYARD HOMESTEADING

Before you make up your mind on whether homesteading is for you, it's a good idea to consider all the benefits. There is no need to uproot your family to become self-sufficient, especially if you can use your backyard to help achieve your dreams. However, if your dream is to live off the grid on a large property, then by all means, have that as one of your large goals to work for as you prepare yourself.

You don't need to start big! Start small. Grow your preferred vegetables instead of buying them from the grocer. Gardening comes with many benefits, both mental and physical. People who have their own vegetable garden:

• take pride in their work

• get exercise, fresh air, and vitamin D

• lower their anxiety and stress

• the elderly strengthen their joints, bones, muscles, mental health, and cognitive function

• can help children's development and help them appreciate where their food comes from

   • creates a bonding and learning opportunity

• improves diet through more variety of produce

- eating better boosts mental health
- will likely lower the number of processed foods consumed

- can give back to the community; when there is excess

For those who want to do a little more than just a home garden, homesteading with livestock also brings many benefits. Depending on the animal, you can harvest meat, milk, eggs, pelts (fur), wool, and even droppings that can be composted for fertilizer to improve the quality of crops.

Becoming self-sufficient allows you to focus on the bigger picture of your life. Caring for your animals and garden is therapeutic and rewarding, allowing you to see the fruits of your labor.

You can improve your self-sufficiency by learning more skills such as preserving food (canning, pickling, etc.), foraging, and trying your hand at beekeeping. Each skill you acquire will give you many benefits, and there are many you can learn.

DEVELOPING THE SELF-SUFFICIENT LIVING MINDSET

One of the biggest limiting factors of homesteading is the mindset. To become self-sufficient, there is a certain way you need to think to make it.

1. Start small.

- There is no reason to move out to the middle of nowhere when your home garden is good enough.

2. Determine what you can and can't live without.

- What can you grow or make yourself?

3. Lower your daily waste.

- Organic waste can be composted to make fertilizer.
- Inorganic waste (cans, boxes, etc.) can be recycled or used in other ways. Get creative.
- In most counties and states, it's legal to harvest rainwater. Use this opportunity to lower your water bill.

4. Start a vegetable garden.

- Choose your favorite plants and determine what will be needed to support them from the seed stage to producing crops.
- Be mindful of hardiness zones and first and last frost days.
- Learn to start plants indoors to get a jumpstart on the growing season.

5. Consider small backyard livestock that isn't difficult to care for.

- Follow up on laws about owning, housing, and sanitation for these animals.

6. Consider which grids you can leave.

- Not all states or counties allow you to leave all the grids (electricity, sanitation, water, etc.). Research your local laws.

7. Lose the debt.

- This is especially true if you want to leave to live on a property while not earning a lot.

8. Lose the addiction.

- The costs to maintain it isn't worth it and will cut into your budget to keep your homestead running.

- A few creature comforts should be worked into the budget.

9. This won't be easy, and life isn't fair.

- Prepare yourself and lose the self-pity attitude. This will be a lot of work.

10. Read!

- Continually educate yourself on how to improve your current homestead setup. Once you have all the research, it becomes easier to make choices and move forward with becoming more self-sufficient.

11. Develop a seasonal to-do list to help you prepare yourself for the yearly work involved in a homestead, no matter how small.

12. Constantly learn new skills.

- Learn to preserve excess food, make your own clothes and cleaning products, fix tools or vehicles, etc.

- The more you learn, the more self-sufficient you can become, within reason.

13. Learn to cook from scratch.

- Fresh ingredients are better for you and have less waste involved with them.

14. Have mental resilience.

- Homesteading isn't for everyone and requires a certain mindset to overcome adversity.
- It's important that you can be independent, care for the environment, and be able to recycle to help reduce waste, can connect with the land you're living on, and be able to live and eat as organically as possible.

Making even one of these changes is enough to shift your mindset to becoming more self-sufficient. Take responsibility for changes you can bring about, and never stop learning new skills to help you make those changes.

UNDERSTANDING THE LEGAL TECHNICALITIES

Homesteading is a legal practice as long as you stay within the scope of the law of the state or county you're in. While you cannot create a homestead on federal land, many states offer initiatives where free land is offered. However, there are often catches involved, usually something like having a home and a working piece of agricultural land within a set period. These pieces of land are usually found in farming communities or small towns. Other potential catches are the regulations you'll have to abide by in terms of zoning and building codes or even deed restrictions. Generally, these places offer up free land to improve their economy.

Urban homesteading isn't without its own rules. Over and above the laws of state and county, there are Homeowners

Associations (HOAs) that may frown upon backyard live-stock and even vegetable gardens.

Before attempting a homestead, be sure to follow up on all local laws. Websites, such as *FindLaw*, are great resources that allow you to look up your state laws surrounding home-steading. It's also a good idea to check local laws as they may differ from state laws.

Before taking up the offer of free land, take note of not only the local laws but also the quality of the land you're getting. Certain non-negotiable resources should be available to you (access to water or roads), or the property isn't worth it. Sometimes something sounds too good to be true because it is.

MYTHBUSTERS FADS ABOUT HOMESTEADING

As with everything in life, there are myths that surround homesteading that need to be busted before you continue with this book.

Accepting free items is helpful	• While you may be keen to accept some tools for free, you never know the quality of what you're getting. They may end up costing you far more in the end. • It's possible to find some gems, but always keep an eye out for quality goods. o This can be applied to seeds and livestock as well.
Can't take holidays	• If you have neighbors or friends willing to look after your property for a few days, you can enjoy short holidays. • The homestead should never be abandoned.
Farmhouse required	• Not at all. • When buying a piece of property, generally, there's no rule that states you have to build a house on it immediately. • There are an array of mobile homes you can use until you want to build a home.
Follow others	• While you should take inspiration from others, you don't need to copy them. • Learn many skills to forge your own way.
Homesteading is a full-time job	• It can be at certain times of the year, especially during harvest time. • You can still hold down a job that doesn't revolve around your homestead. • Homesteading alone may not bring in enough income for you to quit your day job. Although in some cases, your property may allow you to turn a profit.
Homesteading is doomsday prepping	• Perhaps to some, but for most, it's a chance to live sustainably or offset some bills.
Must be organic	• This is a personal preference and not a requirement.
Must get the perfect property	• While the concept is ideal, it likely doesn't exist. • Find which best suits your needs and goals.
Outbuildings required	• As long as the livestock you're rearing have adequate shelter, food, and space, then no barn is required. • Similarly, you don't need to build a root cellar or greenhouse until needed.

Quick to start	• While some goals can be achieved within a year, homesteading takes time to perfect.
Requires a lot of space	• This will depend on the livestock you keep and the goals for your homestead.
Requires property to be rural	• Urban homesteading is very popular and doesn't require you to uproot your family. • Moving to a rural property may be one of your goals eventually.
There's only one right way to homestead	• There are many different types of homesteads and people who develop them. • Developing and running a homestead is as unique as the person doing it.
True homesteading means living off-grid	• Again, this only applies to some people who want to live this way. • If you're interested in urban homesteading, there is no reason to leave the grid unless you want to. Although, it may be difficult depending on where you live.
Will be in harmony with nature	• Not quite. Even with urban homesteading, there's a chance livestock can be taken by predators, ferals, and pets. Even your crops can be devastated by pests, and the weather can turn against you. • You need to find a balance that suits you.

There may even be other myths not listed here. Before you believe them, take the time to research their truth and see if it's something you're willing to believe or not.

ACCESSING NEEDS AND RESOURCES

Every backyard or parcel of land is different in what it can offer you. No one piece of land is the same. However, when starting your homesteading journey, there are some resources that you want available to you, as this will make your journey easier.

1. Consider how much space is needed if using alternative energy sources.

2. Consider the remoteness of any property bought.

- Will impact whether you have access to schools, medical help, and other amenities such as community, stores, municipal services, etc.

3. Depending on what you want from your homestead, you may need a large parcel of land.

- Ensure you calculate proper fencing and housing areas for the different livestock.
- Map out where you want your garden, pasture, orchard, animal areas, etc., to be. This way you can see which are best suited where.
- Place livestock appropriately. Large livestock can be farther from the home, while smaller animals should be closer.

4. If you want to place a home or other outbuildings, you must locate or be able to make level areas.

- Topography is important.

5. The property needs to be accessed easily.

- This needs to be considered for all seasons, as you may need vehicles coming onto the property for deliveries or helping with large projects.

6. It must have easy access to water, or you should own water rights.

- You'll need enough water for you, your family, livestock, and crops.

7. It needs to have sunny spots or areas you can clear to create sunny spots.

- Fruiting vegetables can need up to 16 hours of sunlight to make fruit. Take the time to track the sun through your property to identify which locations are perfect for your crops.
- Preferably, you want your garden south-facing in the northern hemisphere and the opposite in the southern hemisphere.

8. Keep an eye out for potential dangers.

- In rural areas, it is difficult to get help if something goes wrong. You'll need contingency plans for natural disasters, injuries, and illness.

9. Know the perimeter of your property and mark it efficiently.

- Perimeter fencing allows people to know when they stumble onto private property. This also allows you to hang appropriate signage to warn people.

10. Some purchased properties may come with existing infrastructure, this will need to be repaired and maintained or removed if too aged.

11. The quality of soil and water should be tested to see if it matches what is required.

While most of these are for larger homesteads, many of the points can also be applied to backyard homesteads.

IDENTIFY YOUR GOALS AND PRIORITIES

Starting a homestead can be a daunting task if you don't know what you want from it. It doesn't matter whether you're starting a backyard homestead or a large acreage homestead, clear goals and priorities are a must. You might already have a few ideas, so get a notebook and start writing them down, no matter how preposterous they may sound.

While doing that, consider what your priorities are for starting a homestead. Are these priorities aligned with the values you hold dear? If not, it may be difficult to reach your dream. Reconsider some of your priorities, so you can make your goals more achievable. Make a list of priorities and goals, so they can guide you with making short- and long-term goals.

It may be difficult to have goals in place when you aren't sure of the work involved with homesteading. So, if you have no immediate goals you can think of, write down one dream you have for your homestead and consider this a long-term goal. Goals can be divided into 1, 2, 5, and even 10-year goals, depending on what they are.

Once you have your dream written down, you can start working on SMART goals. To understand SMART goals, you need to understand what they mean.

- Make *specific* goals.
- These goals should be *measurable*.
- These goals should be *attainable*.
- All goals should be *realistic* and *relevant* to what you want.
- Goals should be completed in a *timely* fashion.

Now, you can start putting your goals down on paper. Be honest with yourself about how long it will take to complete these goals. However, realize that some goals will be easier to achieve than others, while some will need a little more flexibility because they could be dependent on the seasons, availability of materials, or require money. Goals should be flexible to allow you to feel like you can achieve them. Let's look at a short-term goal and the sub-goals needed to reach it.

Let's say you want a flock of chickens before the first year is completed. Perfect, but you can't have chickens and know nothing about them. Develop a few other goals to help you reach the main one.

- Research what breeds of chickens are available and how to care for them.

- Consider choices such as heritage breeds, dual-purpose birds (used for meat and eggs), chicks versus adults, etc.

• Design and build—or buy—a coop and run to support the number of chickens you want.

• Set up the interior of the coop with the necessities for year-round care.

• Get the necessary food.

• Then, finally, bring your chickens home.

Goals are needed, as they help you focus and make a choice, regardless of the type of homestead you want. While you may want several long-term goals for your homestead, here are a few suggestions for the first year of homesteading.

• Get the garden ready.

  - You'll need to prepare the land or consider raised beds, choose the way you want to grow (vertical, square-foot gardening, etc.), where to have your compost, where to get fertilizer, know about potential pests, and consider what you want to grow.
  - When selecting plants, consider those used in food, companions, and medicinal plants to cover everything you need.
  - Trees should be planted in the first year, as they take several years before producing fruit.

- Add necessary structures and fencing.

  - If you plan to have animals, they need a place to shelter when the weather turns bad. This is non-negotiable.
  - Other potential structures you should keep in mind are greenhouses, as they can be used to grow plants when the weather is poor.
  - Fencing and mesh keep your livestock in and unwanted animals out.
  - Wooden or electrical fencing is perfect for paddocks, while hardware cloth is great for coops and hutches.

- Get the correct livestock for your homestead.

  - There are many laws concerning which animals are allowed in residential zones and in agricultural zones. Be sure to read up on your local laws before getting the animals or they may be confiscated.
  - There are a variety of livestock that you can use on your homestead, such as chickens, rabbits, goats, sheep, pigs, etc. Some are more suited for backyard homesteading.

• Take the time to learn new skills.

  • Food preservation is a great way to make food last through seasons when nothing grows. Meanwhile, foraging can help you find additional resources.
  • Start building a library on homesteading to help you develop the skills and experience needed.

• Consider any tools or machines you may need for developing a homestead.

  • Backyard homesteading will need less equipment than a larger homestead.
  • Always buy quality tools, as they will last longer.

While it's a good idea not to start too big or take on too much responsibility, it's a great idea to get out of your comfort zone to see what you can achieve.

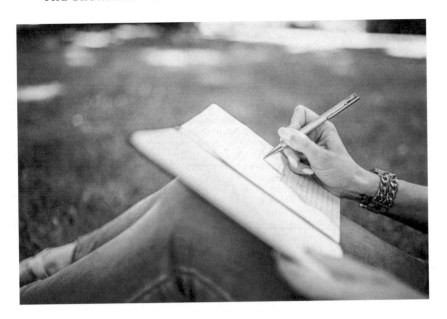

HOW MUCH SPACE DO YOU REALLY NEED?

Realistically, this depends on the goals you have set for yourself and the level of self-sufficiency you want. The closer to being self-sufficient, the more space you'll need. You also need to consider all the people and animals that will live on the property, as they will need their own space and enough space to grow the food they require.

With gardening space, a family of four will need at least 1.8–2 acres of land dedicated to growing vegetables. They will need another 1.8–2 acres if they want to produce grains such as wheat and corn. Then, with animals, they'll need another 3–5 acres (animal dependent) plus an additional 10 acres to grow the food the animals need.

Chickens	• Great for meat, eggs, feathers, and fertilizer. • You'll only need about 8 chickens, which takes up about 200 square feet with their coop and enclosed run. • Scratching in the same place can result in soil erosion, so using a chicken tractor (mobile coop) will allow the ground time to recover.
Goats	• Used for meat, milk, hair, and fertilizer and can clear land. • Two goats only need 500 square feet. • Nanny goats need to be pregnant or have a kid to produce milk.
Sheep	• Perfect for meat, wool, and fertilizer. • An acre of pasture can support 6–10 sheep (breed dependent), but you'll need more land if the pasture is poor quality.
Cattle	• Provides meat, milk, hide, and fertilizer. • Not considered beginner friendly. • Will need two pastures to support growth. Switching between the two will allow them to recover between feeding periods. • A cow will need an acre of pasture and can eat on it well for 80 days before moving to the next pasture.

If they want a functional orchard to give them the fruit they need, they will need ¼–½ acre for about 25 dwarf fruit trees, as these mature faster and take more space than their full-sized varieties. Then, if they want to produce wood for cooking and heating, they will need another 10 acres with trees of suitable size.

Then there are potential outbuildings that need to be built eventually and will need space dedicated to them. These will include a greenhouse or polytunnel (390 square feet), a root cellar for storing food without electricity (8 square feet), and lastly a barn, which can be as big as 1,200 square feet, but acts as storage space and a place to keep animals. This equates to roughly a medium homestead at just over 30 acres. This doesn't even include building a home to live in. However, when there's no need to generate your own wood or animal feed, a family of four could easily survive on a homestead as small as 13 acres.

However, there is no reason to have all these parts to a homestead if you're happy with the urban homesteading experience. With urban homesteading, you require less space, but you won't be able to have everything that can be on a medium-sized homestead. A small family (four members or less) can easily use 0.1 acres for gardening space to help offset the bill for fresh vegetables. A larger family will need at least 0.2 acres dedicated to gardening. With such limited space, planting strategies such as square-foot gardening or vertical growing are the best way to concentrate the land they have to get the maximum yields. If they want to include animals (usually limited to chickens, rabbits, and pygmy goats) they will need an extra three acres, not including acreage to grow their feed. However, there is no need to have these animals, as many urban homesteaders will still get meat, eggs, or milk from the market.

People who wish to be more self-sufficient will need to use their goals to determine the property size they need. This will be dependent on the budget and what's available. Even with thorough planning and mapping, you may not find the property you want within your needs. Be flexible with what you want and what you're willing to give up. Regardless of the property size or backyard you have, be sure you can manage it. Don't be over-ambitious, as this is how something gets overlooked, and the homestead can fail.

BUDGETING IN ADVANCE

If your ideal homestead is several acres of land in another state, it's time to budget, so you can afford to get the land you want. Budgeting before the move will allow you to pay off any debts you may have accumulated and allows you to create a homestead fund for the future move.

The best way to create a budget is to first look at your monthly and yearly expenses. Create a list in a spreadsheet and tally up how much money you're spending monthly and yearly. This is an opportunity to see what you're spending on home expenses (food, clothing, kitchen supplies, etc.), heating, water, rent, credit card payments, vehicle maintenance, etc. Be honest with yourself, even with expenses such as shopping sprees or eating out.

Next, consider how much after-tax money is coming in. Include all investments, earnings, and savings. If you're earning a variable income, always budget on the lowest scale. Add this to the same spreadsheet and see if you're living within your means. You mustn't create more debt than income, as this will continue to bury you under a mountain of debt that is difficult to get out of.

With the numbers in front of you, it's easier to see where you can reduce debts and perhaps generate more income. Start cutting away unnecessary spending, as this will help you with your future budget. Setting up a budget for a homestead is more than when living in a home or apartment. Tools, equipment, livestock and their needs, building materials, and garden expenses are extra payments you wouldn't normally have to pay.

It's a good idea while you're budgeting to look at the cost of items you need for your future homestead. Even if you're only doing backyard homesteading, there are still unexpected costs (livestock getting ill and needing to see an exotic vet) that need to be paid.

Another important part that should be added to your budgeting is an emergency fund. Put a little money in there every month to create a nest egg in case something disastrous were to occur. Budgeting gives you a clear goal to help you plan for the future, so keep track of all your purchases starting today. This is one of the first skills you'll need to learn to achieve your homestead dream.

Don't budget on the homestead making a lot of money quickly. While homesteads can generate some money—especially if you have a remote job that allows you to work from anywhere—there is no guarantee.

PLANNING YOUR HOMESTEADING STYLE

There is no one style for homesteading. Whether deciding how your future house will look or how you plan out your property, everything needs to be according to your goals and budget. Make plenty of sketches of what the ideal property looks like to you. It is with these sketches that you can develop a future budget.

Designate a place for everything you want on your homesteading property. This will include your home (if you don't have one built already), where your orchard will be, where animals will be placed (pastures and paddocks), and your garden or multiple gardening areas. Don't forget to allocate space for potential outbuildings that may be on your long-term goals list. Keep your property size in mind when planning all of this.

The same with planning your future home. This is the place where you'll be spending much of your time turning your crops and animal products into cooked or preserved food. You need adequate space to achieve this. You'll also need to consider your plans, such as children or elderly parents, so there will need to be an adequate number of rooms, or the house will need to be easily adaptable. Sites such as *Homestead House Plans* are a great resource with their customizable home designs to suit all your needs.

Designing Your Backyard Homestead

Make the best of what ground is already in your possession. Take measurements and create a rough sketch that allows you to create a plan. The layout of your homestead will be heavily dependent on the size of the property, what's on the property, and what can be added to the property legally according to local building and zoning laws.

While you can get a lot of inspiration from the internet and books, always keep in mind the size of your property, your goals, and your budget. Even the smallest piece of property can be highly productive when used correctly.

Consider where you'll plant trees carefully, as these could shade out where you want your garden beds to be. The same for later outbuildings such as a greenhouse. You'll also need to carefully consider where your animals will be placed. Small livestock animals don't do well under extreme weather conditions, and their housing will need to not only be well-built but also within a shaded part of your garden. Alternatively, they can be of the tractor design, which allows you to move it where you want it to be.

Don't limit yourself to only growing in a single spot. Divide the area set aside for your garden into several sections and use them for many purposes. This will also help you with crop rotation. Don't forget to calculate walkways, so you can easily move around as you care for your plants. Individual gardens can further be divided to help you plant companion plants and have dedicated trellises for sprawler plants (cucumbers). If you're not ready for multiple garden beds, start with one, and add more as you get more confident.

Your plants and animals don't need to be separated all the time. Permaculture—a way of having your garden being more in harmony with the environment—can be applied to even the smallest homesteads. Chickens can be used to clear up fallen, overripe fruit, and their droppings can be composted to create nitrogen-rich fertilizer for the garden.

THE BEST PLANTING SYSTEMS

Planting is more than just dropping seeds in the ground. Once you have the right place to grow your various plants, it's time to prepare the ground for planting. If your homestead has quality soil, you can plant seeds directly in it, but if the soil is of poor quality or too hard, you'll have to use raised beds.

In terms of large pieces of land being cultivated, there are many ways that seeds can be sowed once the ground has been plowed—unless using no-tilling methods.

• Traditional sowing

  • Seeds are added to a funnel that ends in several points that drop seeds into furrows created.

• Broadcasting

  • Manually or mechanically scattering seeds over a wide area before covering them with a layer of soil, usually with a plank dragged over the surface.

- Dibbling

  - A tool called a "dibbler" is used to make uniform holes in the ground. The seeds are then planted and covered. This is often used in vegetable gardening.

- Drilling

  - Furrows are created mechanically or manually at the correct depth and distance for the seeds being sowed. The seeds are sowed as the furrows are created. Can be done with a plow, a tractor, or bullock-drawn seed drills.

- Transplanting

  - Seeds are grown indoors for a few weeks before being planted outside. Speed planter tools help with planting transplants without digging holes in prepared soil.

It isn't just how you plant your grains and vegetables that is important. How you plan and plant your orchard is also important. Fruit trees should be planted within the first year of your homesteading adventure, as they take some time to mature. While dwarf varieties will produce earlier, they don't last as long as standard fruit trees. Consider planting dwarf and standard-sized fruit trees at the same time. Dwarf trees like to be planted 8–10 feet away from each other, while standard-sized trees prefer 18–20 feet distance.

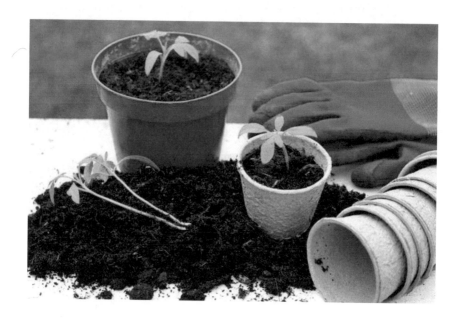

When deciding on fruit trees, always pick what can survive in your hardiness zone and what you like to eat fresh or preserved. Where you place them is also important—especially if you want them to survive their first year. Avoid areas that gather frost or are near buildings, powerlines, or pipes.

Trees can be planted in a freestyle pattern throughout your garden, with no real pattern other than keeping their distance from other trees. Alternatively, they can be planted in rows using square or rectangular systems. With the square system, all the trees are equidistant from each other, while the rectangular system has one set of rows set further apart from others. The spaces between the trees can be used for intercropping herbs and flowers to help attract beneficial insects.

MATERIALS NEEDED FOR HOMESTEADING

There are a variety of materials that will make homesteading significantly easier. While all standard gardening tools will be required—with smaller varieties needed for raised beds—and general tools needed for work around the house, there are a few others that you should consider.

Automatic gate locks and doors.	• Consider getting these to ensure your livestock stays where they're meant to be. • Automatic doors are a great way for chickens to leave their coop without you needing to get up at the crack of dawn.
Broadfork	• Great for loosening compacted soil, especially if you want to go no-till.
Brush hook or machete	• If your property is a little wild, these tools will help you cut back the wild growth far quicker than pruners will.
Bushel baskets and storage bins	• Bushel baskets are great for holding produce as you're collecting it. • Storage bins can be used for anything from storing tools to being used as compost bins.
Compost thermometer	• Hot composting needs to reach a certain temperature to compost correctly. • The thermometer allows you to know when you need to water, turn, or add more organic matter to your compost.
Expanding hoses	• Get water to wherever you may need it.
Fruit harvester	• This tool is a great way to pick fruit without damaging the tree or using a ladder.
Gardening hoe	• A quality hoe will make working large pieces of land much easier.
Mattock and pry bar	• Mattocks are great for breaking through hard soil to prepare it for the growing season or to make holes to bury stakes. • The pry bar can also be used to dig holes, but it can also be used to roll heavy items such as boulders or logs.
Speed planter tool	• Used for quick planting of transplants. • Great if you can't bend down easily.
Zip ties	• Get a lot, as they are a great way to keep something together until you can fix it.

This table isn't a fully inclusive list, but it's enough to get you to start thinking of what you may need. While you always want quality tools, don't discredit estate sales where quality old tools, such as hand planes may be up for grabs. The more self-sufficient you want to be, the more jobs you'll have to be able to do on your property and, therefore, the more tools you'll need.

3

DESIGNING YOUR HOMESTEAD SYSTEM

One of the greatest joys of backyard homesteading is to create a garden layout and rotation system. Once you have identified the perfect place to add your garden or gardens, ensure to map them onto your homestead plan. There are a variety of garden layouts you can try, some of which can be inground or above your existing soil.

Blocks	• This layout is similar to raised beds, where blocks of 3 by 4 feet are made, and plants are grown according to their spatial needs. • This allows for less wasted walkway space but requires fertilizer to grow many plants in a small area.
Four squares	• The gardening area is divided into 4 squares. • Each square is dedicated to plants with different nutritional needs. ◦ Block 1 is for the heavy feeders (corn), Block 2 is for the moderate feeders (peppers), Block 3 is for light feeders (radishes), and the last block is for the nitrogen fixers (legumes such as peas). • Each growing season the blocks move on clockwise so the nitrogen fixers can add nitrogen back into the soil after the heavy feeders. • This layout is a simple form of crop rotation.
Raised beds	• If your soil is too hard or of poor quality, raised beds allow you to garden above them. • This layout can be made with or without a frame.
Rows	• Long rows with ample walking space between them. • This garden layout may have problems with weeds.
Square-foot gardening	• Best applied to raised beds, this technique divides the bed into square-foot sections. • Each square foot will grow a specific plant at a specified density. • The density of seeds is calculated as 12 inches divided by the space required by the adult plant.
Vertical	• Simply put, growing up. This can refer to trellises that allow sprawling plants to grow vertically or stacked containers. ◦ Vertically stacked pyramids with 5-gallon containers are very popular. • When using stacked containers, ensure that there are enough drainage holes.

The main point you need to remember is that gardens should run from south to north, with the tallest plants—or those on trellises—should be planted to the north to prevent them from overshadowing the plants in front of them.

Regardless of which lay you decide to use, it's vital to remember what you planted every year, as you want to prevent a build-up of soil-borne diseases and pests. Keep a notebook with sketches of what you grow yearly, and use markers in your garden as well, as young seedlings all look the same.

PRINCIPLES OF EFFICIENT BED MAKING

Raised beds are the easiest layout to enjoy in a backyard setting, as they can be made from kits or any recyclable material available. There's no digging or tilling required, and poor soil quality won't affect what grows in a raised bed. These beds are so adaptable they can be moved to different areas if need be and can even have automated watering systems built into them if the watering source is too far away.

Raised beds come with a variety of benefits. Being raised, it makes for easier gardening, there is little soil erosion, it drains well, and is easy to apply companion planting and square-foot gardening, not to mention crop rotation. The best thing about raised beds is that it doesn't cost much to make one. All you need is some wooden planks, stone bricks or rocks, or metal sheeting along with screws, a drill, brackets (in some designs), and organic matter with some topsoil.

Setting up a Garden Bed

Ensure the area where you want to place the bed gets adequate sun, is out of the direct wind, and doesn't stand in an area where water gathers. Next, decide on what material you want the bed to be made of. Most people prefer using wood, but you can also use corrugated iron sheets or large rocks. Alternatively, skip the frame completely and use a mound of dirt.

It's a good idea to start with a single bed to see if you can manage it. However, you'll soon have the confidence to add another. Don't forget to add potential locations for other beds to your homestead plan, as you'll need to consider them and the pathways between them. The walkways should be mulched with gravel or wood chips to prevent any weeds from sprouting.

Most raised beds are eight by four feet. While the width should remain four feet, as it's easier to reach into the middle from either side, the length can be as long as you have space. Also consider the depth of the raised bed. A 12-inch-deep bed is generally the best depth, as most root vegetables can grow comfortably without reaching the poor soil below.

Raised beds can be made with many different materials in many different ways. If you're feeling creative, you can build a spiral-raised bed with bricks. Alternatively, you can use containers such as gallon buckets, old milk crates, or even flower boxes to create a raised bed.

Raised beds are also highly adaptable. Adding a hinged hoop over the top allows you to create a shield against pests and the cold, allowing it to act like a greenhouse. You can even add some trellises to the outside of the bed to allow sprawling plants to grow vertically instead of taking up space within the bed. Ensure that the trellises are always placed toward the northern part of the raised bed to prevent tall plants from shading the plants in front of them.

Building a Raised Bed

Preparing an area for a raised bed is as simple as cutting back the grass and weeds and smothering it with cardboard or weed-blocking fabric. You don't need to wait for the plants to die back; you can immediately place the frame.

Get wooden planks cut to the size you want, preferably with a thickness of two inches. You can do this yourself if you have the tools, otherwise, let the hardware store do it for you. When making a large frame, be sure to build it where you want it to be, as some frames will be heavy. Connect them using deck screws, as these are more rust-resistant. You can either connect the planks with screws, or you can use brackets or framing angles to the inside or outside of the frame.

If you want something a little less permanent, you can keep the frame in place with eight two-foot-long pieces of rebar. Add two to each of the shorter ends and four to the longer ends to keep the frame from moving out with the weight of the soil inside the bed.

While you can use recycled wood to make the frame, be wary of pressure-treated wood, as it may contain chemicals that could leach into the soil and be harmful. When purchasing planks, try to get rot-resistant wood such as cedar. While it will last longer, it is more expensive. If you're not averse to replacing your beds more frequently, you can use pine as a cheaper alternative.

You can also use concrete blocks with a hole in the center to make a raised bed. While not as aesthetically pleasing, the central holes are perfect for planting a few herbs or flowers to encourage beneficial insects closer to help with protecting and pollinating the garden.

PRINCIPLES BEHIND VERTICAL GARDENING

The reason people like to grow vertically is because it allows less ground to be utilized for the maximum yield possible. Not only this, but growing plants up off the ground lowers the chance of soil-borne pests and diseases from getting to the fruit. By growing vertically, more of the plant is exposed to sunlight and has less chance of developing moisture-related diseases, such as powdery mildew, as there is more space for air movement.

There are a variety of vertical gardens available, some as complex as the hydroponic towers to as simple as the plant wall or biowall. Some work on a system of having water and nutrients pumped toward the rooms, while with others the material the plants are in is made of a porous material containing all the nutrients the plant needs, and water can be added. Many kits can be used with gray water collected from the home or can have a timer installed to water the plants periodically.

If you have spare wall space and would like to add some greenery, vertical growing bags are perfect for small greens and herbs. These may need more water than other plants as walls often create a rainshadow, and they won't get water naturally. Vertical gardening is dependent on the size of the plant you're growing. The more robust the vertical growing garden, the easier it is to grow heavier crops.

Raised Bed Versus Vertical Gardening

While this may seem like a difficult choice to make, it comes down to what you like and have space for.

Raised bed	Vertical gardening
• Great for deep root vegetables such as potatoes. • Perfect for people who can't bend down. • The raised bed warms quicker in the spring. • It drains well. • These beds can be built to specifications. • Can control how much water is used.	• Large containers allow for lots of room to grow. • More yield using less space. • Using drip irrigation means less water wastage. • Difficult for large animals to get to crops. • Some vertical systems are moveable. • These beds are usually made of rot- and rust-resistant materials.

While hydroponic vertical towers have become popular lately, trellises, arches, and even A-frames are also a form of vertical growing and can easily be combined with raised beds. The heavier the fruit the plant grows on a vertical structure, the stronger it should be. While teepee structures are perfect for beans and peas, they can't support melons, whereas arches can be made from cattle gates and can support the weight of melons.

INSTALLING IRRIGATION, FENCING, AND POWER SYSTEMS

When living on a homestead, you're responsible for many aspects of the land. In many cases, you may have to educate yourself on how to install different systems such as irrigation, lighting, and even fencing.

Power

Many alternative power sources can be used on a homestead. Solar power is the most commonly used, but if in a suitable area, you can also use water and wind turbines to generate power. However, before you get excited about leaving the power grid, you'll need to know how much power you are using monthly. Get a wattage meter to see how much power your electrical devices are drawing, and how much each uses in a day. Then you can determine not only how much you need to budget for power, but what items you don't need to waste electricity on. With a power budget in place, you can determine how many batteries and solar panels are required to generate the power to run your homestead. Always get a little more than you need, as this will help on days when there isn't enough sun. More on this in Chapter 7.

Irrigation

The best irrigation to use is drip irrigation, as it prevents wastage, delivers water where it's needed, and there is less chance of weeds spreading, as they aren't getting watered. For container gardens (pots, buckets, etc.), one of the best watering methods is a soda bottle with a hole in the lid with its end cut off. Depending on the size and number of holes, the water from the bottle will slowly leak into the container, watering the plant. All you need to do is to top up the bottle now and again.

However, this method doesn't work on a larger scale, but drip irrigation still works. A hose can be connected to a series of PVC piping that goes toward your garden. At the start of the garden, the PVC piping can be connected to irrigation piping (garden piping) that splits toward the various rows of plants. The irrigation pipes closest to the plants can have holes punched into them and have drip emitters added to them to allow water to drip at the base of the plant. While the drip emitters aren't needed, it does allow the water to get closer to the plant. These systems can even be connected to a timer!

Fencing

There is a variety of fencing you can use, but before you do, you need to do some research as to what the fencing is meant to do. Is it meant to keep animals out or in? If it's designed to keep animals in, you must understand the nature of those animals. Pigs can't jump, so they don't need a tall fence. However, chickens can fly over, so they would need taller fences or an enclosed area.

A fence is only as strong as its line and anchor posts. If these aren't stable, the whole fence can be pulled over. Where you put the gates is also important, as you want to have easy access to them. Then, before putting a fence in place, know your zoning laws as well as where your property lines are!

The type of fencing you use is important. Hutches and coops can be made as an enclosed area that prevents the animals from getting out and predators from getting in. Generally, these structures use hardware cloth with various-sized holes. The smaller the hole, the less likely a predator can get in, but the more difficult it is to cut.

For larger animals, you're more spoiled for choice. Woven wire is great for keeping sheep and goats in place, while the barbed wire is better for cattle as they can't push the fence over. High tensile non-electric fencing is good for horses and mules but not much else. The lightweight electrified poly wire or tape is inexpensive to install, and some varieties are known to keep predators at bay but cannot be used for large perimeters. Each type of fencing material has pros and cons that should be researched before installation.

PART II

FOOD PRODUCTION AND PRESERVATION

In this section, we'll cover how to produce different kinds of food and how to preserve it.

4

GROWING YOUR OWN FOOD

For a homesteader that wants to grow directly in the soil, they may find the task a little more difficult than those who grow in raised beds as they contain amended soil. Most soils today have been leached of nutrients needed by the plants, have too much clay, may be contaminated, or maybe too compacted to dig into. Thankfully, there are ways to breathe new life into that soil without damaging it further.

No-tilling methods allow you to grow plants without damaging the soil structure or removing what nutrients it has left. Using methods like the Back to Eden Garden and the Lasagna method allows a homesteader to grow above the current soil they have.

How to	
Back to Eden	• Add a layer of newspaper to the soil and soak it. • Add an organic layer (compost, soil, lawn clippings), followed by wood chips, and then a thin layer of manure. 　○ Green manure (fresh) can burn plants, so spread it thinly.
Lasagna	• Add a layer of cardboard or newspaper and wet it thoroughly. • Add alternating layers of green and brown material until you create a 2-foot-thick layer.

These methods do add a layer you can grow in and return nutrients to the soil, but it won't be instantaneous. Building soil health can take years. Even if you use raised beds, never leave them with no plants during winter. Grow a cover crop such as clover or alfalfa to protect the soil. Once winter has passed, these plants can be worked back into the soil to add nutrients.

SAVING SEEDS

Saving seeds is important for a homesteader who doesn't want to waste money buying seeds yearly. While not a crucial skill, it's one worth learning, as it helps create stronger plants, and you may even develop heirloom seeds to hand down to your children.

Saving seeds is as easy as planting them, and this is where you need to take notice of what seeds you're using. Heirloom and open-pollinated plants have the seeds you want to save, as these are strong plants that will produce offspring that resemble the adults. While hybrid seeds can produce higher-yield crops or crops resistant to certain diseases and insects, their offspring may not look like the adults as the genes are spread through the different seeds.

There are different methods of collecting seeds, depending on what plant you're dealing with.

Lettuce	• Allow lettuce to go to seed (bolt). • Collect the fluffy white seeds, add them to a paper bag, and then shake. • Remove the chaff and collect the seeds at the bottom of the bag.
Peas and beans	• Allow pods to dry on the vine until they can be easily popped open. • Remove seeds and continue to dry until a fingernail cannot leave an impression when applied.
Peppers	• Allow the fruit to wrinkle on the bush before picking. • Remove the seeds and allow them to dry.
Tomatoes	• Tomatoes are a little more complicated, as the seeds need to ferment in some water for a few days. • Scoop the seeds out and add them to a jar of water. • Cover the lid with some mesh to prevent pests from getting in and stir the contents now and again. • After 2–3 days, some seeds will start to float. Remove and discard them. • Drain the rest of the contents through a fine sieve and wash the seeds. • Allow the seeds to dry.

Some plants won't make seeds within their first year of life, generally making some in the second year. These types of plants are known as 'biennials.' They will need to overwinter under thick mulch, and once spring comes around, they will continue to grow and produce seeds.

All seeds should be stored in paper envelopes that are well labeled with the type of seed and when they were harvested. These packets need to be kept away from moisture, direct sunlight, and heat. Some stored seeds can last decades, while others will only last a few years.

How to Choose Viable Seeds

When using stored seeds, there is no guarantee that they will still be viable and germinate. While many people believe that seeds can be stirred into water and the floating seeds aren't viable. However, this method isn't foolproof, and it's better to use the paper towel method to check for viability. All you need are your seeds, two pieces of paper towel, and a sealable bag.

1. Take 10 seeds and place them between the paper towels.
2. Wet the paper towels and pat them together before adding them to the baggie.
3. Seal the baggie and place it in a warm area. Most seeds like to germinate at 70 °F.
4. Keep an eye on the baggie and spray more water if it appears the paper towels are drying out.
5. Germination depends on what seeds you used, but most seeds should germinate between 7–10 days.
6. Open the baggie and check the seeds.

Using 10 seeds allows you to determine the estimated viability of the stored seeds. If all 10 seeds germinated, you'll likely have 100% germination. If there are only 8, then the germination will likely only be 80%. By the time you reach

50% viability, you'll need to sow double the seeds to get the same estimated number of plants as seeds that have 100% viability. Below 50% germination means the seeds aren't worth sowing.

However, keep in mind that some seeds need specific conditions for them to germinate successfully. Some need scarification (the hull needs to be scoured to allow moisture in) or stratification (where the seed needs to go through a period of moisture and cold to simulate winter).

FRUIT TREES AND BERRY BUSHES

At this point, you already know the perfect location for your favorite fruit trees. Don't try to grow a fruit tree or berry bush from a seed. It takes too long, and there's no guarantee it will grow to adulthood. It's best to go to a nursery—or order online—a plant that's one to two years old. This gives you a jump start on your plant's production, as it can take some time before they can produce a decent crop.

Fruit trees		Berry bushes	
Apple	2–5 years	Blueberries	1–3 years
Apricot	2–5 years	Currants	2–3 years
Citrus	1–3 years	Honeyberry	1–4 years
Peach	2–4 years	Raspberry	1–2 years
Pomegranate	2–3 years	Strawberry	1–2 years

The duration represented in the table is how many years extra the plant will need to grow before it starts producing.

Some of these plants cannot be planted by themselves and will need a second to help them cross-pollinate for greater yields. Not only that, but trees and bushes need to be pruned and trimmed yearly to allow better growth.

Choosing the right plant will come down to your hardiness zone (as identified by the USDA Plant Hardiness Zone Map) and how many chill hours (hours under 45 °F) it requires. The distance trees need to be planted from each other depends on the variety they are, and berry bushes should be at least 50 feet away from trees to prevent them from being shaded or their roots competing for water.

How many trees and bushes are planted is dependent on what you're growing.

Mature dwarf apple	5–6 bushels (40–48 gallons)
Mature dwarf pear	6–8 bushels
Blackberry	35–70 cups
Blueberry	15–45 cups

For trees, you can get away with 1–2 for the whole family, but with berries, you'll need 2–4 plants per person to get enough of that particular fruit. This is why variety is so important, or you'll get bored with what you're eating.

CULTIVATING MUSHROOMS

If you're interested in growing mushrooms, it's fairly simple and can be done indoors or outside. First, you need to start with some edible mushrooms. These could be button mushrooms from the store or kits (with spores or plugs). Other edible mushrooms you can use include, but aren't limited to, oyster mushrooms, lion's mane, chicken of the woods, and shiitake to name a few. Mushrooms have preferred growing media. Button mushrooms enjoy compost, while oysters like straw, and shiitake grow well on hardwood sawdust. Each is unique to their needs, so do your research when selecting a mushroom to grow.

To grow indoors, you'll need a bucket or an old laundry basket that's been thoroughly cleaned, some growing medium, and some mushrooms or a kit.

1. If using mushrooms, place the gills of the cap down on some wax paper and leave them undisturbed for 24 hours. This will allow the spores to fall onto the paper.
2. Once the spores are ready, or if using a kit, fill the container with a growing medium that's been moistened.
3. Add the spores and cover the container with some cling film that has holes in it. This will allow for some airflow but keep most of the moisture trapped.
4. Keep the container at roughly 60–80 °F.

5. Harvest mushrooms once the caps separate from the stems, this can take 10–15 days, depending on the variety grown.

Growing outdoors is a little more complicated and will take more time. Select logs of either hardwood or softwood. Turkey tail grows well in all types of wood, and lion's mane enjoys chestnut, oysters like birch, and chicken of the woods enjoys maple.

1. Create spawning logs by drilling 1 ¼-inch deep holes and adding the mushroom plugs to them.
2. Knock the plugs into the wood and cover the hole lightly with some soy wax.
3. Place logs in a shaded area with good ventilation where they can absorb moisture well. These logs should be watered weekly at least, more often if it's hot.
4. Eventually, the budding mushrooms will start to grow, but it can take several months, depending on what you want to grow.

An alternative to growing outside is adding spores to some soaked cardboard and covering them thickly with straw before wetting them now and again to keep them moist.

When growing mushrooms outside, ensure to pick only those you planted, as not all mushrooms are edible.

HOW TO CULTIVATE HERBS AND MEDICINAL PLANTS

Living on a homestead far away from doctors can be a slight problem if you don't have an apothecary of plants ready to treat simple ailments such as headaches or sunburn. Luckily, there are an array of herbs and medicinal plants that you can grow.

Arnica	• An external muscle rub can be made with this herb together with coconut oil or beeswax to help with sprains and sore muscles.
Basil	• Great for food but can also be used to treat stomach ailments such as gas, diarrhea, and constipation.
Feverfew	• Can be used to make a wonderful tea to fight headaches and migraines.
German chamomile	• Can help fight fevers and aids in relaxation.
Lavender	• A wonderful flower that attracts pollinators, which can also be used to lower anxiety and treat headaches, insomnia, and even depression.
Lemon balm	• Tasty in ice teas, as well as helping fight heartburn and indigestion. • This herb can also help to lower stress and anxiety while fighting insomnia.

These few plants only scratch the surface of what you should be planting in your garden. Many herbs and medicinal plants are also considered companions to many vegetables, making them perfect for integrated pest management. Basil is known to repel tomato hornworms, protecting tomatoes and making them taste better.

You don't even need to plant them together with your vegetables. You can create a garden specifically for medicinal care. Select your favorites or delve into determining the various pros and cons of the different plants to see if they suit what you may need. Learn how to harvest, dry, and create different teas, ointments, etc., for any ailments your family may suffer. You can even cultivate a vertical wall of herbs.

A warning though; plants such as stinging nettles and mints —which can be used to treat many ailments—are known to spread like wildfire. It's best to keep these in containers, so you don't have to deal with their invasiveness.

INTEGRATED PEST MANAGEMENT

Integrated pest management (IPM) is an approach where you try to bring harmony between your garden and the pests that would eat it. There will always be pests, but how you react to them will determine whether they remain or get worse. Pesticides have been used for decades to deal with pests, but these poisons aren't always used correctly, resulting in insects becoming resistant. There are five ways IPM is implemented.

• Prevention

  • Keeping plants strong prevents them from being targets for insects.

• Sanitation or cultural practices

  • Clean up areas that may host pest species.
  • Practice polyculture (growing different crops), crop rotation, and companion planting to deal with pests.
  • Improve soil quality by adding organic matter to return nutrients.

• Mechanical

  • Remove insects when noted. Hand-picking insects and eggs help prevent their numbers from growing.
  • Take the time to go through the garden to inspect it.

• Biological

  • Many beneficial insects feed on or parasitize pest species. Lady bugs, praying mantises, spiders, and parasitic wasps can be attracted to a high-diversity garden.

- Chemical

  - Always the final resort. When using chemical sprays, be sure to read and follow the instructions.
  - Use targeted pesticides that get rid of insects causing problems.
  - Consider using natural pesticides like neem oil or insecticidal soaps first.

For IPM to work, it's vital to correctly identify friends from foes and what kind of damage the foes can do. Even if your garden does take damage, develop a threshold of what you're willing to suffer before reacting severely. There's no reason to throw an apple away because of one worm, it can still make a fine jam once the worm is disposed of.

IMPROVING YOUR PLANTS' IMMUNITY AND RESILIENCE

If you want to prevent your plants from being attacked, you need to start by strengthening them. A strong plant is more resilient to attack than one that is stressed. You can help strengthen your plants above and below ground through foliar (leaf) and plant-strengthening fertilizers added to the soil. These can help enrich the soil as well as increase the absorption rate of the plant and make the soil microbes happy, many of which can help the plant grow better.

One of the best plants to make these kinds of fertilizers from is aloe vera. This plant is known to enhance plant resilience, promote vigor, and boost healthy growth. The preferred species is *Aloe barbadensis*, as it's edible and has medicinal qualities. While *A. chinensis* can also be used, it isn't edible and should only be used as a soil drench and not a leaf fertilizer.

To make a soil drench fertilizer, you'll need ¼–½ cup of aloe leaves to a gallon of water. Blend the leaves to create a smoothie-like consistency and pour it into the water. This mixture is best applied in the mornings, within 20 minutes of it being made. Apply ½–2 cups per plant, poured at the base, preferably after they have been watered. Powdered aloe can also be used, but only use ⅛ teaspoon per gallon of water.

To make the leaf fertilizer, the same measurements are used, but the leaves need to be skinned first, or the insides must be scooped out. Doing this will prevent the mixture from clogging the nozzle you're using. Shake well and use it soon after making it, preferably spraying in the morning.

ADDRESSING COMMON GARDEN PLANTS DISEASES

If you have a garden, it's only a matter of time before a disease will affect your plants, no matter how hard you work. While there are thousands of diseases—some specific to certain plants—there are a few that are more common, and you should learn to recognize them.

The best way to treat your plants is to correctly identify the problem affecting them and get the correct treatment plan. There are other ways to lower the chance of disease affecting your plants.

- Lower plant stress by dealing with insect attacks, not planting too early, and giving ample water when hot.
- Rotate crops to lower the chance of diseases building up in the soil.
- Avoid overcrowding and overhead watering, as this promotes fungal growth on leaves.

- Add mulch, as water splashing back from the soil can spread diseases.
- Remove and destroy infected plants or infected parts.
- Add balanced fertilizer to keep plants strong.

Bacterial canker or blight	<ul><li>Usually attacks stone fruits such as peaches, plums, etc.</li><li>The fungus is spread through rain, as it causes the spores to travel to new locations.</li><li>Fruit develops black sunken spots.</li><li>Leaves develop black spots, which rot through leaving holes.</li><li>This disease can also affect blossoms, buds, and twigs.</li><li>The best treatment is to trim away the infected plant material, as much as 16 inches. Do this when the plant is dry to prevent spores from spreading. Then, treat with copper-based fungicide.</li></ul>
Botrytis blight	<ul><li>Also known as "gray mold."</li><li>A gray mold develops on dead or dying parts of plants and is spread due to high moisture and overcrowding.</li><li>Clear away dead material and increase air circulation.</li></ul>
Powdery mildew	<ul><li>Causes downy mold to develop on leaves.</li><li>This is caused by high moisture, humidity, and overcrowding.</li><li>Remove affected sections of plants and thin out the weaker plants.</li><li>Fungal treatments can help.</li></ul>
Rust	<ul><li>Small red, brown, or orange spots develop on the underside of leaves. More unsightly than damaging.</li><li>Remove the affected areas and treat them with sulfur or copper-based fungicides.</li></ul>
Verticillium wilt	<ul><li>Not all diseases can be treated, and this is one of the worst.</li><li>The roots are affected, leading the plants to become yellow and wilted. In some cases, one side of the plant may experience young branches dying.</li><li>Trees may show black or brown streaks under the bark.</li><li>There is no cure, and the plant needs to be removed and destroyed.</li><li>It's best not to plant anything in that location for a few years.</li><li>Any tools used should be thoroughly sterilized.</li></ul>

However, if your plant is suffering from a disease, it needs to be dealt with quickly to prevent it from spreading. While there is little that can be done with bacterial infections; to treat fungus, there are several treatment plans.

- There are natural controls such as *Trichoderma harzianum, Bacillus subtilis, Streptomyces griseoviridis,* and *S. lydicus* are sold under many trade names.
- You can apply neem oil or copper- or sulfur-based fungicides.
- Other possibilities include bicarbonates such as potassium bicarbonate, ammonium bicarbonate, and sodium bicarbonate—though the latter will need to be in neem oil to be effective.

Regardless of the treatment you're trying, ensure that the instructions are followed.

5

RAISING ANIMALS FOR FOOD
AND COMPANIONSHIP

U nless you're a vegan, having animals on your homestead will not only provide you with different foods but also functional animals that have some kind of purpose. Before considering *any* animal, consult your local laws and HOAs to see if you can have agricultural animals on your property. You may find that permits are needed to own certain animals or to slaughter some of those animals.

Every animal on your homestead must bring something to the table, whether it's a food source, or it works for you. These animals need to be adaptable to the climate you live in, and you'll need to take care of all their needs from food, water, shelter, and veterinary care. Remember that many agricultural animals count as exotic pets, and the vet fees may be high.

Some of the best animals to include in a homestead include honeybees, meat rabbits, chickens, pigs, sheep, goats, and cows—all depending on the size of your property. The larger the animal, the more resources it'll need. Likely in an urban setting, you'll only be able to care for rabbits, chickens, and possibly honeybees.

PRINCIPLES OF ANIMAL CARE

Animals are living organisms, and their well-being is your responsibility. You must do the extra research to ensure that any homestead animal in your care receives all the requirements for a happy, stress-free life, regardless of them being reared for the pot.

Each animal is unique in its needs for food, water, shelter, and medical care. Ensure that the animals you have are never in need of the first three, and when the fourth is needed, act quickly and accordingly.

All animals should be treated with care and dignity. If you cannot or are unwilling to slaughter animals, have them taken to a professional who can do the deed without causing undue stress, fear, and pain.

Housing and shelters for all animals are something that needs to be constantly monitored and updated to prevent animals from escaping, remaining sheltered, and preventing predators from getting in.

HOW TO CARE FOR HOMESTEAD ANIMALS

With an array of backyard-friendly livestock, and those best suited for larger homesteads, you're spoiled for choice when it comes to picking the animals that will join you.

Cattle	<ul><li>Per cow, an acre of grassland needs to be made available and rotated every few months to allow it to recover.</li><li>It's important to remember that large animals can be a danger when not handled daily.<ul><li>Respect the flight zone (the area around the animal that makes it nervous), and never chase after a startled cow. Allow it to calm down before approaching it.</li></ul></li><li>Milk cows will need to be bred to produce milk, and even then, the milk supply will eventually dry up.</li><li>Males tend to be aggressive and serve no purpose on a homestead other than meat. If this is the case, ensure that bulls are made into steers (castrated).</li><li>Most homesteads only have a place for one cow, but multiple are possible depending on property size.</li></ul>
Chickens	<ul><li>Each chicken needs 3–4 square feet on the coop floor and 10 square feet in the enclosed run.<ul><li>If free roaming, the run space isn't needed. However, be wary of predators.</li></ul></li><li>Will require a coop with roosting bars (10 inches per bird) off the floor, nesting boxes, and enrichment toys.</li><li>The run should include a scratching or sand bathing area and access to food and water.</li><li>Breed selection is important.<ul><li>There are three types of chickens; meat, layers, and dual-purpose.</li><li>One of the best dual-purpose is the black Australorp which can lay about 250 eggs a year, with roosters weighing 8 ½–10 pounds and hens at 6 ½–8 pounds.</li></ul></li><li>May not be allowed to have a rooster in an urban setting, but they aren't needed for egg production.<ul><li>Will be needed if you want chicks.</li></ul></li><li>Chickens don't have teeth and will need oyster shell grit or diatomaceous earth with their feed to help break it down.<ul><li>They can also enjoy limited amounts of kitchen scraps and cracked corn.</li></ul></li><li>Chickens can overheat easily. So, their coops need to be in the shade and well-ventilated during the summer. During the winter, insulation can be added to keep the heat in.</li><li>Check food and water daily.</li></ul>

Goats	<ul><li>Urban homesteads will likely only allow dwarf or pygmy goats.<ul><li>These goats need at least 10 square feet of stall space.</li></ul></li><li>Large goats will need up to 25 square feet of stall space.</li><li>An acre of land is enough to support 2–8 goats depending on the quality of the land. Luckily, goats share space with sheep and cattle well.</li><li>Two goats are enough to produce milk and dairy products for a small family.</li><li>A three-sided shelter is more than good enough for a goat if you don't have a barn.</li><li>Hooves need to be trimmed every 6–8 weeks, so get young goats used to being handled.</li><li>Require fresh food and water daily and will need milking up to twice a day.</li><li>The pen area will need to have sturdy fencing at least 4 feet high. While wooden paneling is fine, be wary of goats who get their heads trapped between slats.</li><li>Bucks are considered aggressive and likely won't be allowed in urban settings. If you want your nanny goats to produce milk, they will need to be taken to a farm that offers studding (breeding) services.</li></ul>
Horses	<ul><li>Require a minimum of 1 ½–2 acres of good pasture or a minimum of 4,500 square feet to exercise in.<ul><li>Some horses may need up to 4 acres, especially when breeding and spending up to 80% in the pasture.</li></ul></li><li>Depending on the size of the horse, stalls can be 10 by 10 feet or 12 by 12 feet.</li><li>Most homesteads won't have a use for a horse unless they pull a plow. However, with enough property, you can board horses.<ul><li>This allows owners to have a safe place to keep their horses.</li><li>Additional zoning laws may limit the number of horses that are allowed.</li><li>May require additional rotational fields.</li><li>Hobby farm horse boarders can be 5–10 acres and have 5–6 stalls.</li></ul></li><li>Boarding horses is very expensive and requires a lot of skills and experience to attempt.</li></ul>
Pigs	<ul><li>Require 6–8 square feet indoors.</li><li>Outside runs should allow for 20–50 square feet per pig.</li><li>Pigs should always be in pairs due to their social nature.</li><li>Fencing needs to be sturdy, as pigs are strong!<ul><li>Hog panel fences will keep them in place.</li></ul></li></ul>

	o Pigs are great escape artists, so if they escape, don't chase them! It'll result in them running and screaming. Lure them back with some food.
	• A shelter can either be an A-frame or a three-sided shed.
	• Pigs need a lot of food. Aim to feed quality feed with extra protein to improve meat quality.
	   o Pigs are also happy to have kitchen scraps, clear old vegetable patches, and clean orchards.
	• Pigs can have a strong odor, therefore, clean their pens weekly and hand a few charcoal bags to help absorb the smell.
	• They will require a lot of water and a muddy patch to wallow.
	   o Mud protects their skin from the sun and parasites.

Rabbits	• Within a hutch, rabbits need ¾ square feet per pound of mature weight.
	• The hutch needs to be easy to clean and well-ventilated.
	   o Rabbit tractors are an easy way to allow your rabbits to forage in different places on your property.
	• Females need access to a nesting box when pregnant.
	• Preferably, you want meat rabbits, unless you intend to show them, then show rabbits are fine.
	• You only need to start with a male and a female to build up your numbers. Afterward, only allow males and females together when you want to add rabbits to your freezer.
	   o Always take the female to the male for mating as they are highly territorial and may attack the male if it's brought to them.
	• Food and water should be checked daily, and the hutch cleaned weekly.
	• These are quiet animals, and if you keep up with cleaning their hutch, your neighbors won't even know they're there.

Sheep	• Sheep will need roughly 15 square feet of stall space, especially if a ewe has lambs. • An acre of land is enough to support 3 sheep. • Require the same type of shelter and fencing as goats. • The feed should mostly be grass and hay with very little grain. • Do well with shallow water troughs kept in the shade to keep it cool and algae-free. • The pen needs weekly cleaning to prevent parasites and fly strikes. • Similarly to goats, they need their hooves trimmed but also their teeth checked, as they can grow very sharp and cut tongues as they chew. • Depending on the breed, a sheep should be sheared 1–2 times a year. o Shearing is a skill worth learning.

All animals will benefit from vet visits and getting all necessary vaccines and deworming—especially goats, sheep, and pigs. Coops and hutches should be raised above the ground to prevent predators from digging into them.

Other potential animals you can keep on a large homestead include those that are considered guard animals. While you may think this job usually is left to dogs, there are an array of other animals that can also serve in this role. Donkeys are known for territorial behavior and will bray when danger is close. Even animals such as peacocks are an early warning system if something is amiss.

BEEKEEPING

Bees are considered one of the most important pollinators worldwide, and they are in danger of dying due to colony collapse disorder (likely caused by parasites) and foulbrood. To have these pollinators on site to pollinate your plants is invaluable! However, beekeeping or apiculture requires that you educate yourself on what these insects need to thrive. A single hive can contain thousands of bees, all being commanded by a queen to do a variety of jobs from nursing to collecting honey.

If you're lucky enough to have a swarm on your property already, all you need then is a hive and a permit to keep them, though do check your local laws for more information. If you don't have a local swarm, you'll need to get one. Thankfully, bees can be purchased as a package (a queen with bees who aren't related to her) or a nuc (a nucleus colony where a hive is already partially created in frames). It's best to order bees in the winter, so they can be ready in the spring for the flowers.

Once your bees are ordered, you should purchase equipment such as protective gear, the smoker, hives, frame lifters, and a bee brush. When looking to get equipment ensure it is perfectly clean, as you don't want to infect your hive with diseases or parasites.

Next, set up the hive in an area that will have plenty of flowers close by. The hive must be protected, as this will keep the bees alive longer. It should be in a sunny location, but shaded from high afternoon temperatures, sheltered from the wind, and in an area where it can stay dry. Ensure the entrance to the hive is pointed away from high-traffic areas, as you don't want to walk through where the bees are flying.

Once the hive is set up, all you have to do is wait for your bees to arrive. Once they do, check on the queen, without her, the colony won't make it. Allow her into the hive first, and the rest of the hive will follow soon after.

If there are no flowers about yet, you'll need to provide them with some sugar water to tide them over until pollen is available. Ensure this is close by in a shallow dish, so they don't drown. The new hive should be checked every one to two weeks to see if the bees are settling and the queen has started laying eggs. This is a true sign that the hive is functioning well.

Winter can be particularly hard on bees. To help them survive, the entrance to their hive and ventilation should be limited but not completely closed. Instead of sugar water as a supplement, granulated sugar should be offered. Depending on how a hive is designed, you can place a feeder within the hive, or add the sugar to some paper towel placed on top of the frames. Don't allow the paper towel to block off any of the ventilation. You can add 1–2 cups of sugar on top of the paper towel.

These insects aren't only great for pollination, but they also produce a wide range of products such as honey, beeswax, propolis, and even royal jelly. Two hives are enough to give you up to 30 pounds of honey a year, which is a great replacement for sugar, something you would have to purchase when on a homestead. Beeswax and propolis are great for your home apothecary. Beeswax is an ingredient used in muscle rubs, ointments, and even candles. Propolis comes from plant sap, and the bees use it to keep their hives clean and disease free. Propolis can then be used in tinctures and infusions to give the same benefits to people. With so many wonderful items being produced by bees, these are animals you can't overlook having on your homestead.

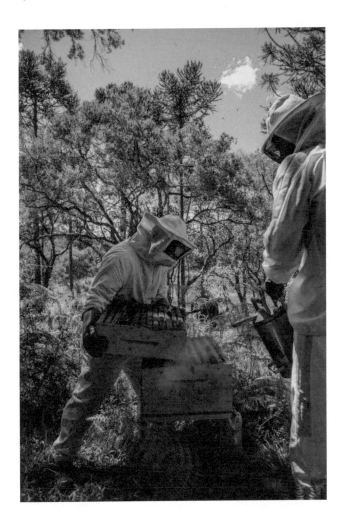

6

FOOD PRESERVATION TECHNIQUES

Once your crops start coming in, you'll soon realize there is no way you can eat everything harvested within a few days. Thankfully, there are many different ways that you can preserve and save food for future meals, especially during seasons when nothing, or very little, can grow. This is one of the first skills you should learn on your homestead. Some preservation methods are easier than others, while some require electricity or specialty tools. Before deciding which method is best for you, consider the cost of buying equipment (if you don't own it already), space requirements, and how long the food is preserved for. Regardless of which method you decide to use, ensure that you are always labeling what the food is and when it was preserved.

DRYING

Drying or dehydrating food is one of the oldest techniques used by people around the world to preserve food. The basic principle is to remove the liquid from vegetables, meats, herbs, and fruits to prevent bacterial decay. Most meats to make jerky will also be salted or added to a brining solution before being dried to ensure the bacteria are killed.

Drying is done in three ways; sun drying, with a dehydrator, or oven drying. While sun drying doesn't require any power, it takes the longest to dry and is subjected to the weather.

The pros to this method are that the food becomes light-weight and easier to store. It can also be coupled with freezing or vacuum sealing. The cons are that nutrients (particularly water-soluble nutrients) are lost, and depending on what method you use, can use a lot of electricity. The food also needs to remain in an airtight container to prevent moisture from ruining the food with mold growth.

Before dehydrating, herbs, fruit, and vegetables need to be washed and then patted dry with paper towels. Fruits and vegetables need to be cut to 1/4–1/2-inch thick; if too thick, the pieces will need more time to dry. Fruit and vegetables can also be dipped into lemon juice to reduce browning as they dry. Preferably, you want to slice meat thinner if you want to dehydrate it to jerky. The pieces are then added in a single layer to a tray and dried using the sun, dehydrator, or oven. While sun drying can take days, a dehydrator set to

135 °F needs 6–8 hours, while an oven set to 200 °F will take 2–3 hours.

FREEZING

Freezing is considered another easy way to preserve food for later use. This can be applied to all types of food—including ready-made meals—as long as they are frozen once they reach room temperature.

The advantages of freezing are that it doesn't require a lot of labor and is easy to do. The disadvantage is that some foods need blanching to help lock in the nutrients, flavor, and textures before freezing. Even then, there are some nutrients lost during the freezing process. Other than that, you're reliant on power to keep this food edible and need a lot of space.

If you're concerned about possible power outages, you can add a cup of frozen water with a coin on top of it to monitor your freezer. As long as the coin remains on top of the ice and not frozen inside of it; the freezer has maintained its integrity.

Most food can last six months to several years, especially if you vacuum seal the food before freezing. Food being frozen must be done so in airtight containers to prevent ice crystals from developing on or in the food. Freezer burn is something that occurs the longer food is in the freezer. While this doesn't make food inedible, it doesn't taste very good.

Food should be kept in airtight, freezer-safe containers or even wrapped in freezer paper. Always check your freezer space before deciding to add anything to it. Never over-fill your freezer, as this can lead to power air circulation and the food not freezing uniformly. Avoid fluctuating temperatures by opening and closing the freezer doors. Know what you want before you open the doors.

CANNING

There are two ways to can food. The first is the water bath canning method, and the second is the pressure canning method. While pressure canning requires a pressure canner, it is the safest way to preserve low-acidic foods such as meats, vegetables, soups, fish, and seafood.

Water bath canning is done with foods that fall under a pH of 4.6, as the temperature reached is only 212 °F. This is perfect for most fruits (though some will require some lemon juice to increase acidity), pickled vegetables, condiments, vinegar, salsas, and chutneys. It's important to keep the acidity of the food in mind, as foods with a pH higher than 4.6 need to be pressure canned or there is a risk of botulism.

Pressure canning raises the temperature of the food in jars to over 240 °F, effectively killing even the botulism bacteria. However, how you use your pressure canner will depend on your elevation above sea level, so be sure to read the instruction manual. Recipes used for pressure canning need to be USDA-approved and come with strict instructions on how long you need to cook at your elevation.

You'll know your canning was successful, once you remove the jars from the canner and set them aside to cool. While cooling, the lids create a seal, keeping the food safe. Any popped lids (not sealed) should be added to the fridge and eaten within three days. These jars can be reprocessed, but the contents tend to become mushier than intended.

The advantages of canning are that the food is shelf-stable, doesn't require refrigeration, and retains most nutrients. Pressure canned foods can last one to two years. The disadvantage is that you must know the acidity of foods to determine which is safe to be water bath canned or pressure canned. The equipment isn't too expensive, but new canning jar lids need to be used every time, and any chipped mason jars should be disposed of. The mason jars take up a lot of space, so be prepared to have a place for all the jars or a root cellar with plenty of space. Once opened, these jars should remain in the fridge and be eaten within a few days.

Some extra equipment that will make canning easier include the magnetic lid holder, tongs, and a wide-mouth funnel.

FERMENTING

There are three types of fermenting:

- ethanol fermentation

  - Uses yeast to convert different carbohydrates into alcohol and carbon dioxide.
  - Will need to buy a starter culture or get some of an existing culture.
  - Different starters will create different foods.
  - Common ethanol-fermented foods include sourdough, kimchi, sauerkraut, and beer.

- lacto-fermentation

  - Also known as 'lactic acid fermentation.'
  - The *Lactobacillus* bacteria are used to convert different carbohydrates into lactic acid.
  - This fermentation is used to turn milk into cheese and yogurt, as well as make pearl onions, pickled vegetables (no vinegar involved), and olives.

- acetic acid fermentation

  - This fermentation will require a SCOBY (symbiotic combination of bacteria and yeasts).
  - After alcohol is made with ethanol fermentation, oxidation of the alcohol creates acetic acid.

- This fermentation is used to make vinegar.

The advantages of fermenting are that it's a safe way to preserve your food, it's good for your gut as it introduces good microorganisms, and you don't need any specialized equipment to do it except a fermenting starter. The disadvantage is that fermented food is often an acquired taste, and regardless of what you make, it needs to go into cold storage eventually.

Regardless of what you want to ferment, it's important to follow recipes closely to prevent mistakes and problems. Fermentation causes a build-up of gasses, and in some cases, if jars aren't burped (opened a little during fermentation), they could explode.

Dairy Products

Milk is best stored in the fridge, but it won't last long, so other preservation methods should be employed. Milk can be used to make butter, yogurt, cheese, and buttermilk. All of these need to undergo some fermentation to reach their final products. Ensure that you read your recipes well to know exactly what you need beforehand, so you aren't halfway through a recipe and don't have what you need. This will only lead to waste.

Making Vinegar

If you have some overripe fruits, these are perfect for making vinegar and don't have to end up in the compost. To make vinegar, oxidation is needed (the addition of oxygen) during the fermenting process. You'll also need a starter of some kind. One of the easiest to use is raw apple cider vinegar that comes with the mother. The mother is a gel-like substance that is found on top of the vinegar that contains the acetic acid bacteria. These are the organisms that will help turn your fruit into vinegar. How sour the vinegar is will depend on the sweetness of the fruit used. You can use most fruits, such as apples and berries, to make vinegar.

Chop the desired fruit into smaller pieces and add them to a jar before adding the water, sugar, and starter with the mother. Quantities will depend on the recipes used. Stir the contents until the sugar has completely dissolved. Add a cloth as a lid and tie it in place with a rubber band.

Allow the mixture to ferment at room temperature for three weeks, stirring at least twice a day. You'll note bubbles starting to form in the liquid at this time. After three weeks, the solids can be strained out, and the liquid can continue to ferment with the cloth lid for another six weeks. A new mother will develop on top of your vinegar, and you can scrape it off to make more vinegar.

After six weeks, the vinegar is ready and can be placed in a fresh bottle and sealed. It's important to use a sealable bottle, as the fermentation process using oxygen needs to be stopped or the vinegar will spoil. Store vinegar at room temperature.

While you're tempted to use this homemade vinegar to preserve foods with pickling, don't. Commercially available vinegar has a pH of 2.4 (5% vinegar concentration), so unless your homemade vinegar can reach this pH, it cannot be used as a preservative but rather a tasty addition to your food.

PICKLING

Pickling is often confused with fermenting due to some of the similarities between the two and how pickles can be created with both methods. The main difference is that pickling uses salt and vinegar to preserve foods while fermenting specifically uses microorganisms.

Often referred to as 'quick pickling,' this preservation method can be used on eggs, fish, vegetables, fruits, and even meat. The salt and vinegar help remove the water from the food, stopping bacterial growth. Some pickled goods can then be further preserved using the pressure canner. However, note that eggs cannot be pressure canned safely, even when pickled.

CURING MEATS

There are many ways to cure meats depending on your taste preference and what equipment is available. The three most common curing methods are drying, smoking, and salting. Some common cured foods include pastrami, bacon, salami, and even the more exotic foods such as hákarl (fermented and cured Greenland shark) and mettwurst. Curing helps to preserve taste and texture.

Smoking

When smoking, you can decide to do a hot or cold smoke. Cold smoking only uses smoke to preserve the food while adding unique tastes. Hot smoking cooks the food and preserves it with smoke. Hardwoods are best to use when smoking meat as softwoods tend to make the food bitter. Smoking food adds to its flavors and preserves meat, thanks to its antimicrobial nature.

You can create a smoking shed by adding some bars to a shed and hanging meat over it to be cured by smoke. The pieces high above the fire will be cold smoked, while those closer will be hot smoked. Alternatively, a tarp wrapped around a drying rack with meat can achieve the same effect. Keep the temperature of the area being smoked to 109–160 °F depending on the type of smoking technique you're doing.

Salting

Thin slices of meat—fish included—can be layered with salt to draw out all the water. By the end of the process, which can take a few days, the meat should look and feel like leather. The meat should then be placed in airtight containers or frozen to increase its longevity.

Drying

Drying meat can be done in a dehydrator but can also be done in a well-ventilated room. The meat can be left in salt, or have salt rubbed over it to partially dry it out, before it is hung up to continue drying. This process is used to make biltong and jerky. While the meat is hanging, it needs to be protected from insects, as they will still be attracted to the meat. Some people like to dry their meat all the way through, while others don't mind it being a little more damp. However, be wary that moist meat will stand a bigger chance of hosting mold.

JAMS, JELLIES, AND SYRUPS

Jellies and jams are mostly made with fruits cooked together with sugar until a thickened gel consistency is reached. Jams generally contain pieces of fruit, while jellies are made with fruit juice.

When making jams, jellies, and some syrups, you may need to add some powdered or liquid pectin. This chemical—which occurs naturally in all fruit—is required to help with the smooth and semi-solid consistency of these products. It also allows the cooking time to be reduced. However, that doesn't mean you have to use pectin; it just depends on the recipe that you're using. Recipes with fruits such as apricots, strawberries, peaches, and raspberries will almost always need the addition of pectin.

The fruit or fruit juice is then cooked with lots of sugar—with some or no pectin—until, when dripped from a spoon, it slides off as a sheet and not drops. At this stage, the liquid is added to waiting warm jars, and the lids are added, tightened only with the fingertips. If you want to preserve jams and jellies for longer, use mason jars and put them through the water bath canner according to your elevation.

Syrups are made similarly to jellies but with less pectin, as they should remain liquid. Usually, the fruit is cooked with just enough water to cover it. While cooking, the fruit is mashed to release as much juice as possible. Once the fruit is made into a pulp and the cooking time is completed, the contents of the pot can be put through a strainer, the finer the better (tea towel). If you want a clear syrup, allow the liquid to slowly run through the cloth. Alternatively, the towel can be wrung to get as much juice out as possible.

The juice is then added to a new pot with sugar and cooked further. The recipe may or may not call for pectin; follow it exactly. Once complete, the fluid can be added to a jar and then stored in the fridge or preserved further with water bath canning.

DRY AGING FOOD

Dry aging is simply controlling the rate of decay of meat. It's the use of cool, circulating air over an extended period to preserve the texture and flavor of the meat. Doing this allows the meat to become more tender as the enzymes break it down.

All you need to do this is a fridge, a sheet pan, and a wire rack. Take a piece of muscle meat and generously cover it in salt or a brine solution. Don't add any oils at this stage, as it will prevent the meat from absorbing the salt. Once coated well, add the meat to the rack over the sheet pan and place it close to the fridge's fan where it can get even airflow over it. The meat will drip throughout the process, and it can take 10–14 days for the meat to be dry-aged. Once fully aged, the meat can then be cooked or sliced thinly and served raw.

PART III

RUNNING YOUR HOMESTEAD

Now that you have the basics of how to start a homestead and preserve food for the future, it's time to look at how you'll run your homestead and keep it afloat.

7

ENERGY, WATER, AND WASTE MANAGEMENT

W hether you want to be an urban homesteader or someone who wants to be off-grid, it's important to take control of producing energy, saving water, and dealing with waste that will pile up around your property. The more self-sufficient you want to be, the more responsibility you'll have.

RENEWABLE ENERGY SOURCES

For energy requirements, many people prefer to remain grid-tied (where they still have electricity from the grid) or want to be more self-sufficient and go off-grid, not relying on any power produced by the grid. Many people no longer trust the grid, as it has become unreliable, uses too many fossil fuels (nonrenewable energy sources), and can be expensive due to mining and transportation. Thankfully, there are a variety of renewable energy resources that are used to generate electricity for your homestead.

Geothermal	<ul><li>While not a new renewable energy source, it's becoming better understood.</li><li>Geothermal power can be used to generate heat, keep the house cool, or produce electricity.</li><li>To keep the house to the preferred temperatures, pipes filled with coolant or water are buried within the soil below the house. A geothermal heat pump is used to circulate the water.<ul><li>The temperature difference between the soil and the ambient outside temperatures is always different, which helps regulate the temperature within the home.</li><li>On warm days, the ground is cooler, therefore cooling the house. The opposite is true for winter.</li></ul></li><li>To generate electricity, you'll need to be in an area where geothermal activity is present.<ul><li>Drilling into the ground to get to the heat, which is used to create steam. This steam is then used to power a turbine to generate year-round power.</li></ul></li></ul>
Solar	<ul><li>Solar panels absorb the energy from the sun, which is then converted for use on the homestead.</li><li>Power can be used as is on the homestead, or it can be stored in batteries, so you can have power during times of no energy production.<ul><li>The type of batteries you use is important, as car batteries don't cut it.</li><li>You'll need deep cycling lead acid batteries, as they last longer.</li><li>Quality batteries are expensive, and you'll need a few to create a battery bank. This allows you to have energy, even if the system is down.</li></ul></li><li>When using batteries with a solar system, ensure that you also have a charge controller which controls the charging of batteries and prevents them from being overcharged.</li><li>Location is key, as areas that are south facing or are hot or dry tend to do very well with generating energy using solar power.</li><li>Solar is considered the most effective renewable energy source.</li><li>You'll need to ensure panels aren't affected by shade.<ul><li>Cloudy or short days won't generate as much power as cloudless summer days.</li></ul></li></ul>

Water turbine	• This is similar to the wind turbine, as this turbine generates energy mechanically as water runs across the blades of the turbine.
	• For this to be a viable energy producer, you'll need to install the turbine in an area with running water.
	  ○ While this can be a constant way to produce power, winter, and drought can affect the effectiveness of this system.
	  ○ Despite this, you have a better chance of generating more power from a water turbine than a wind turbine.
	• One of the drawbacks is trying to move the power generated from the water turbine where it's needed. There needs to be extensive piping put in place to either move the water to generate the power closer or wires to send the power to where you want it.
	  ○ More wires and pipes increase the chance of something going wrong.

Wind turbine	• For wind turbines to give you the necessary energy, they need to be in a location that gets consistent wind throughout the year.
	  ○ The lay of the land can help wind be generated. Areas such as the mouth of a canyon or the top of a hill are ideal places for wind to occur.
	• Before deciding on a wind turbine, the land needs to be monitored for a few months to determine if the turbine can generate enough power to keep up with your needs.
	  ○ The average speed needs to be no less than 10 miles per hour.
	• Unlike solar energy, which is collected passively, wind turbines generate energy through movement, resulting in more wear and tear.
	• You'll likely need a permit to have a wind turbine, and your neighbors may not enjoy you having one in an urban setting.

While all these options are plausible, they may not all be viable with the kind of homestead you have or your power budget. Other potential sources of power, which should be used as backups, are a diesel generator (can be pricey), a car inverter (allows you to use the car's power to power your house), and extra batteries in case of emergencies.

As discussed in Chapter 3, it's important to know what electronic devices are drawing power and how much, as this will determine how much power your homestead needs. While tempted to have a drier, heater, or aircon in your home, these electronic devices pull too much electricity, and you should probably avoid them.

That said, there are other ways to keep a home warm or cool. Wood or propane heaters and stoves are great to heat the home and cook food at a fraction of the cost of what the cost of electricity will be needed for the same job. Geothermal systems can also climate control your home, but that's not your only option. How a house is built can also have an impact on how it can stay warm or cool. South-facing homes are warmer in the northern hemisphere (especially during winter), and with well-placed windows, opening a few will create a throughflow. So, if you intend to build a home, think of how you can take advantage of nature.

Regardless of which renewable energy source you decide to use, make sure that it suits your needs and what your homestead can give you. Do thorough research and look at what other homesteaders are doing in your area. Don't forget to follow up on local ordinances before deciding anything. The last thing you want to do is put something in place and you have to tear it down later due to a law you didn't know of.

RAINWATER HARVESTING AND GRAY WATER RECYCLING SYSTEMS

On a homestead, you'll be using a lot of water. Cooking, cleaning, and watering your animals and garden are just some of the tasks where you'll need to use water. Water is a resource that no one can afford to waste, as there is only so much freshwater available for use. There are two ways you can save water; rainwater and gray water. However, before you get excited, you must understand that there are some regulations with these saving methods.

Rainwater Harvesting

Not all states in America allow you to harvest rainwater without heavy regulation. States such as Arkansas, Colorado,

Illinois, Nebraska, and Utah all have limitations on how much you're allowed to harvest, and you'll need permits. Many other states don't have as strict regulations, and some even encourage people to harvest rainwater to have non potable water for home use.

If you're in a state that gives you the freedom to collect rainwater, take the time to determine how much water you can collect in a year. The calculation to use is:

The surface of the runoff area can be any roof that water can run from into gutters and then be collected by pipes to go into storage tanks or rain barrels. It's never a good idea to collect rain from the ground, as there are more contaminants in this water than on roofs. The total liters is just an estimation, so be prepared to have a little less due to some evaporation and droplets bouncing from the roof. However, if you're in an area that is high in pollution, there is a chance that the water collected can be contaminated and will need cleaning.

While the total rainfall is important for your plans of storage, it isn't a good idea to buy a tank large enough for annual water rainfall. Do research on when most of the rain will fall in an area and determine how many tanks or the size of the storage containers are needed to have the rainwater in.

Untreated rainwater isn't *safe* for drinking, bathing, or cooking with, as it's considered non-potable. However, if you're living on a homestead with no running water, this water should go through a series of filters and disinfection (bleach, boiling, etc.) steps to make it safer. If not in a situation where you don't have access to fresh running water, this water is perfect for irrigation, flushing toilets, or washing clothes.

Harvesting rainwater is dependent on how much rainfall you have, and even then, it's an average. There's a chance that you may have to have water delivered if your homestead requires it and there are no other water sources you can use.

Gray Water Harvesting

Gray water is water generated when food or clothes are washed, or water from showers, sinks, and baths. This doesn't include toilet water, which contains human waste (urine, feces, etc.) and is considered 'black water,' which is unsuited for immediate recycling.

Gray water cannot be made safe to drink, and should only be used for toilets, washing clothes, and watering trees or shrubbery. Many US states don't allow gray water usage in irrigation. On a homestead, gray water can be added to plants, but only if it doesn't touch the parts of the plant that will be eaten, and even then, it's a risk as the soaps and detergents used can have an effect on the pH of the soil and the microbes within it. If in a situation where gray water is needed for watering plants, avoid harsh chemical soaps and detergents and aim to use biodegradable and environmentally friendly varieties.

When installing tanks for recycling gray water and rainwater, ensure they don't mix at any stage. While many tanks are easy to install, if you want tanks with filters and pumps, you may need a little more help.

WASTE REDUCTION METHODS

Between 30–40% of the food produced in America goes to waste in landfills (Deanna, 2020). While food does get old, and you may no longer want to eat it, there are ways to use that food to improve the quality of your homestead. Lower your wastage by creating compost or using vermiculture.

Composting

As a homesteader—backyard or otherwise—you cannot turn your nose up at compost. Considered 'black gold' by all that use it, compost is a sure way to give your growing plants all the nutrients they need to thrive. Not only does it improve the health and quality of your plants, but it also improves the health of the soil and the microorganisms that live in it. In addition, compost doesn't burn your plants as some chemical fertilizers can when applied incorrectly. Compost can easily be made from the comfort of your kitchen or outside —as long as you can prevent it from smelling!

Compost is the decayed remains of green and brown materials collected from the kitchen or garden. Green materials (high in nitrogen) are fresh garden waste, kitchen scraps, and even manure, while brown materials (high in carbon) are dried leaves, newspaper, and cardboard.

Compost bins can be purchased relatively cheaply but can also be homemade. For home use, you need no more than a 5-gallon bucket with a few holes drilled into the top. The lid should be lockable and have a fine mesh inside to prevent insects from getting in. This lid can also have a built-in carbon filter to help with possible smells.

Once the setup is done, you layer the brown and green material in 3–4-inch layers. If too dry, add a little water; if too wet, add more brown material. Turn the mixture every few days until a dark, crumbly, earthy-smelling material is produced. This can take some time. Storage bins or trash cans can be used similarly. If you have animals prone to breaking into composting areas, purchase containers that can be locked and can't be broken into.

However, if you want something more substantial, you'll need to create a three-crate compost system outside. While there are many ways to build this system, the easiest and cheapest way is to have seven pallets, some screws, and some chicken wire.

1. Connect three of the pallets along their edges to create the back wall.
2. Take two pallets and connect them to the ends of the wall to complete the frame of the system.

3. Add the remaining two pallets to create the three crates within the frame.

4. Staple the chicken wire to the inside of each of the crates.

5. You can also use some extra wood paneling to block the front of the crates to prevent the compost from spilling out, but this isn't necessary.

Now that you have the system created, it's time to use it.

1. In the first crate, layer the brown and green material until the crate is full. This can take some time.

2. Move the compost into the second crate and turn it every couple of days, monitoring the moisture level.

- Some people don't turn the compost at this stage and allow the materials in the first crate to reduce by half before moving into the second crate. This is a more passive approach.

3. Any new material should be added to the third crate during this time.

4. By the time the third crate is full, the second crate should have started turning into compost.

5. A sieve can be used to sieve the contents of the second crate into the first crate. The dark, earthy material will fall

through the sieve, with the larger pieces of non-decomposed material being held back.

6. Store or use the compost and start a new layer of compost on the first crate while turning the contents of crate three into the second crate.

7. Continue until no more compost can be made.

An alternative to this system is the heap system. This is when the layers are added and wetted before turning it now and again.

The crate or bin system is a type of passive composting (unless you're turning the compost) where you don't have to do much to get the material to decay. However, this type of composting can take months or years to reach the point that it's usable in the garden. Luckily other composting methods can be faster.

Burying	• This method is used to dispose of kitchen waste without attracting unnecessary pests. • Trenches are dug close to rows of plants, and the kitchen scraps are buried. • The scraps attract soil microbes and worms to decay the organic matter, releasing the nutrients to the plants growing close by. • This method should be used sparingly, as it can also attract pests and animals to the burial sites.

Compost tumbler	• This is a smaller scale of passive composting.
	• The turnable container allows you to turn the compost with minimal effort, allowing the addition of extra oxygen to the mixture.
	   o There is less smell, critters can't get in, and there is a balance with moisture.
	   o Generally, the compost decomposes better in this environment than passive composting.

Composting in place	• This method is where you chop up the garden waste and allow it to compost in one place.
	• This method is best suited to garden waste and not food scraps.
	• After chopping or cutting the garden waste smaller, it should be spread over resting beds to protect the soil as a mulch.
	   o As the organic matter rots, the nutrients leach into the soil.
	   o The remains of the mulch can then be worked into the soil once spring arrives.

Hot composting	• This method of composting can be a little tricky, as you need the correct moisture, composition, temperature, and volume to achieve your goals.
	• What causes the decomposition is the microbial action of various microorganisms.
	   o Compost can be created in 3–8 weeks.
	• Taking 2–4 parts brown material to each green part, this compost can be built up to 4 feet high and 4 feet across.
	   o Always cover the green material with brown to prevent smell and soak up the extra liquid.

<table>
<tr><td></td><td>

- o An alternative is 1 part food scraps, 3 parts brown material, and 1 part green material, all added as a 4-inch layer.
- A compost accelerator can be added to assist decomposition.
  - o These are a combination of fungi and bacteria that can be purchased or made.
  - o Animal manure or human urine can be used as natural compost accelerators, or you can make a homemade accelerator.
  - o All you need is a gallon of warm water, a can of warm, flat beer, a can of warm, flat soda, and ½ a cup of household ammonia. Mix in a 5-gallon container and pour over the compost.
    - ▪ Ensure the soda and beer aren't light or sugar-free.
- The compost is working at its best when it is moist, and the temperature is 130–160 °F. Any higher or lower than this, and the microbial action will slow down, resulting in halted decay. Turn the compost to introduce more oxygen.
  - o Be wary that composts exceeding 160 °F are at risk of catching fire. Keep an eye on the temperature and spread out until cooled sufficiently.
- A lot of maintenance and resources are needed for this composting method to work.
  - o You'll need to monitor the temperature, turn when it's too wet, and moisten when too dry.

</td></tr>
</table>

Not everything is compostable. While items such as eggshells (crushed up), tea leaves (some bags aren't biodegradable), and even coffee grounds can be added to a composting system, foods high in fat, meat, dairy, and even bones shouldn't be added to the compost as they will only attract pests. Other items you should never add to your compost are human waste (except some urine as an accelerator), pet waste, dead animals, treated plant material or wood, diseased plants, or any weeds that are flowering or making seeds. Why waste money buying fertilizer when compost is just as good and cheaper?

Vermiculture

Vermiculture is another form of composting where worms—annelids such as *Eisenia fetida,* otherwise known as "red wigglers" or "compost worms"—are used to decompose organic matter. Some containers are small enough to be used in apartments, while larger ones can be placed outside (15–55 gallons). While the worms themselves don't do much other than eat what you give them, it's what they leave behind that's so nutritious for your garden. Worm castings (worm feces) are high in bioavailable nutrients—meaning the nutrients can easily be absorbed by plants—as they act the same as slow-release fertilizer granules. Some nutrients they can return to the soil include phosphates, calcium, magnesium, and nitrogen. They are also perfect for your soil, as they assist in aeration and drainage while helping the soil retain more water.

Making a worm bin is as easy as making a compost bin. All you need to start with is something as simple as a storage bin (solid color) with a locking lid with no holes in it.

• Drill ¼-inch holes along the side of the container to allow fresh air in.

• Place the bin where the temperature doesn't fluctuate beyond 55–85 °F. Beyond these temperatures will result in worms not working as well.

- Temperatures beyond 35 °F and 95 °F will kill the worms.

• Now, the bedding needs to be added. This isn't food, but it is used to add grit to the worms' diet, as they have no teeth to chew the organic matter that will be added.

- Bedding can be hydrated coco coir (moist, not wet), shredded newspaper or cardboard, or some soil, hay, or straw.
- Lay down 3–4 inches of bedding and moisten it.

• Order your worms online. Depending on the weight of what you order, you can have 500–2,000 worms.

- Consider how many worms you need according to the size of the container you have. While the population can double within 90 days, it's normally limited by the size of the container they're in.
- Add the worms to the container with bedding by digging a hole and adding them to it before covering them.
- Add a layer of moist newspaper over the top and allow them to settle for a few days.

  ■ When happy, the worms will want to remain in the container and not try to crawl out. If they are trying to crawl out, then the moisture and oxygen levels aren't correct.

• Once the worms have had time to settle, you can start to feed them.

  • Some good examples are vegetable and fruit scraps, young and green garden trimmings, spent coffee grounds or tea leaves, crushed eggshells, and even some sourdough starter—not too much, as this adds more liquid to the system.
  • Fluff the contents of the container before digging a hole to add the food and then cover it again.
  • With every second feeding, add a handful of more bedding as the original bedding will start breaking down.
  • While the worms can eat their body weight in food, this is only under ideal conditions. Never overfeed worms, as this can lead to uneaten food decaying.

  ▪ Depending on the numbers, feed once a week to three times a week.

  • Keep the container moist and sprinkle with water if it gets too dry. If too wet, add some new bedding.

• To harvest the worm castings in the container, instead of adding food to the center of the container, add it closer to the side to encourage the worms to feed along one edge. This way, you can collect the composted material on the opposite side.

- It's best to wait a few weeks to allow the worm castings to build up.
- If you gather a few worms up, that's not a problem. They will do well in raised beds and the garden.
- The worm castings can be added to the top of the soil before working it in.

There are many ways to make worm bins, but this is the simplest of designs. However, don't feel limited to it, as some designs can help make worm tea, which is a byproduct of the worms' activity and is perfect for your garden or indoor plants.

OFF-GRID LIVING: CHALLENGES AND REWARDS

Living a self-sufficient lifestyle isn't for everyone, as it comes with many challenges. The more self-sufficient you want to be, the more challenging it will be. However, regardless of the type of homesteader you want to be, some rewards match the challenges.

Challenges

The initial start-up of getting a homestead ready can be expensive, especially if you want to move completely off-grid and be responsible for your own energy needs and waste removal. This sort of lifestyle can be isolating, and it's not just about distance. While living on a homestead far away from social interaction can be isolating, living the homesteading lifestyle can also be isolating, as not everyone is as excited as you about your latest hen or that the apple harvest is looking good.

The biggest challenge around homesteading is the legality of it. No one law covers the homesteading lifestyle—urban or off-grid. Each state has laws, and these laws can further be different within the state according to local laws. It takes a lot of research to know exactly what you can and can't do, and even then, there's a chance that you can still be fined over an obscure law you didn't know about.

Another annoying challenge is that the land you may find when planning to go off-grid doesn't have the resources required to envision your dream. You may still have to rely on others to bring you water or remove waste—especially human waste if you don't plan to install a septic tank or use compost toilets. You may find that you will have to change the way you think and do things to adapt to having access to fewer resources.

You're in charge of everything that needs to get done. If you don't do it, no backup will do it for you. You're responsible for growing and managing your own food, generating energy, and taking care of your livestock. It's a lot of responsibility, and you need the mental strength to do this daily.

Other challenges you may want to consider are if you need to build a new home and if you have the necessary skills to do maintenance on everything on your homestead, ranging from home repairs to energy-creating systems, vehicles, and even tools.

Rewards

Yes, there are a lot of challenges involved in homesteading, but it isn't all difficult. While the startup of getting your homestead running can be expensive, you can save money (or at least put a dent) in the cost of food and energy in the long run. You can become independent of many things that tie people in place. You don't have to rely on others to do things, as your homesteading skills are learned. You're in control of your life with no other boss but yourself. You decide how self-sufficient you want to become, whether urban homesteading or off-the-grid living.

One of the best rewards is being able to connect more with nature. While you may not always be in harmony with nature, there's a lot to be learned watching the seasons and how your homestead changes with them. Knowing what changes are coming will allow you to use it to your benefit.

Being self-sufficient helps you to lower your carbon footprint. Even something as small as growing vegetables in your backyard is enough to lower the number of trips you have to take to the grocery store to pick up vegetables. Less travel means less fossil fuels used. Even composting reduces the waste you produce, plus you're creating a substance that helps put nutrients back into the soil!

While homesteading can be a simpler way of life, it requires a lot of work to make it simple. Only take on as much responsibility as you can afford to. There is never any reason to move off-grid immediately when you can take your time building up a nest egg and learning valuable skills that would benefit you when you do eventually make the move.

8

STAYING AFLOAT

Well done, you've done it. You have started your homestead journey and now have a running homestead in some capacity. Now, all you need to do is to keep it running. To keep a homestead running smoothly, it's a good idea to draw up a maintenance checklist of things you should be doing with every season.

Here are some activities you should be completing per season.

Spring	• Clean all outbuildings and animal enclosures. • Determine a planting schedule and start transplants. • Do all repair work to buildings, garden beds, and enclosures.

Summer	• Know the order of harvest, so you can prepare to preserve.
	• Plan large-scale cleaning of animal enclosures.
	  ◦ This is more than weekly cleaning.
	• Clear gutters of debris to prevent a leaky roof.
	  ◦ This can also be done in spring, depending on when the rainy season is.

Fall	• Check all animal enclosures and shelters.
	  ◦ Repair and seal if necessary.
	• Do all winterizing necessary for the homestead.
	• Prune bushes and trees, especially those that have already lost their leaves.
	  ◦ Trees that are still harvestable during fall should be pruned in the spring.

Winter	• Clear out the chimney and ensure you have enough firewood to last. • Clean and store all tools, as this prevents rust. • A great time to work on indoor chores such as preserving food and repairing clothing.

This is only a fraction of the work the homestead will need, but it gives you a great start in grouping work according to seasons.

Winter is a great time to reflect on what you have achieved and what you should plan for the next season. It's also a great time to review your budget. Go through all the receipts accumulated during the year and determine if you remained within budget or if you need to change something. Decide what worked, and where cuts can be made for a better financial year.

Everything within the home that is being preserved or saved should be well labeled so that everyone knows what it is and how long it has been there. Come together as a family to do this, as it's pointless to have one person know where everything is. What happens if something happens to that person? The homestead can't stop functioning.

Even after all the planning you have made or will be making, realize that not everything will go according to your plan. Learn to be flexible enough to handle anything that goes awry.

WINTERIZING YOUR HOMESTEAD

As the weather cools and the last harvest is collected, it's time to think of winterizing your homestead. If you're an urban homesteader, this only means protecting your raised beds, covering your compost, and ensuring your animals are protected against the cold. However, as an off-grid homesteader, there is a lot more work that needs to be done. While states with mild winters will have less to do, those with snow for an extended period need a prepared homestead.

Preparing the animals

- Move worm bins to warmer locations.
- Winterproof beehives.
  - Build a windbreak if needed.
  - Hives can be wrapped according to the practices of the state, as each winter is different.
  - Have an upper entrance that bees can use as needed and allow extra ventilation.
- Have all your animals checked by a vet and ensure they are all up to date with their vaccinations.
- Winterize shelters by draping runs with plastic, checking for drafts in the shelters (but still allowing enough ventilation), and adding insulation where needed.
  - You can also add some extra bedding.
  - In some cases, you may have to resort to heating. If so, do it safely, as you don't want to start a fire.
- Check the water frequently to ensure it isn't icing over.
  - Move troughs or water containers to somewhere warm or float something in them.
- Have emergency food and water enough for all animals for at least three days.

Preparing the garden

- Plant garlic, as it likes to overwinter in the ground.
- Bring in the last of the harvest and preserve.
  - Some plants, such as carrots, peas, and even collard greens, can experience some frost before harvesting.
  - Check the frost tolerance of plants before risking them remaining out with the frost.
- Clear garden beds and cover in mulch or plant cover crops.
- Trim all perennial herbs (echinacea, feverfew, lemon balm, etc.), usually up to six inches, and place thick mulch around them.
  - Alternatively, they can be dug up and placed in pots until springtime.
- Annual herbs will need to be harvested.
- Berries with canes will need to be pruned and mulched. Strawberries can be covered with mulch and left in the soil.
- Trees can also be mulched.
  - Add four short stakes around the tree. Wrap burlap sacks around these stakes and fill them with dried leaves as mulch.
  - Younger trees benefit from having a pest-proof layer wrapped around their trunks before adding mulch, protecting them from gnawing pests.
- Remove irrigation systems or use an air compressor to clear them of any standing water.
- Remove temporary stakes and trellises.
- Clear the garden of any fallen leaves and long grass.
- Cover the compost area with a thick layer of plastic or straw.
- Take several soil samples to be tested for soil health.

Preparing the homestead
• Check the owner's manual for any equipment and vehicles for the correct way to deal with them during winter. • Clean tools and put them away. You can even add some mineral oil to them to prevent rust.
Preparing yourself and home
• Ensure you have the correct winter clothing. • Keep tools for shoveling snow in your home.   ○ Easier access than leaving them in a shed outside. • Check vehicles for functionality and safety. • Watch the weather closely and listen for warnings. • Block drafts within your home, as it'll be easier to keep it warm. • Protect exposed pipes with newspaper to prevent them from bursting. • Have emergency food and water stowed away for at least three days. • Have an emergency kit ready in case of disasters.

Creating a list of what you need to do may seem overwhelming, but it's the best way to prevent something from being forgotten. Looking after your animals should always be your first concern.

TRADING AND BARTERING HOMESTEAD GOODS AND SERVICES

While bartering and trading may not play a large role in urban homesteading, it can be the lifeblood of some off-grid homesteads. Inevitably, there will always be something you want from someone, and if you're lucky, they may want something you have, allowing you to complete a trade. Money may still exchange hands now and again, but many homesteaders will be happier making a trade than a sale. As long as both parties are happy with the fair trade, bartering can encourage communication and cooperation between all in the community. When bartering—no matter how fair or honest the other party may be—always keep track of what to whom you're trading. It's also a good idea to have a time limit for the trade that needs to be done—especially when one item is handed over while waiting for the return offer.

It isn't just goods that can be traded. Many homesteaders will happily trade items and goods they have for someone willing to help on their homesteads or have someone with skills such as woodwork or plumbing. Regardless of what kind of homesteader you are, join like-minded people to make a group willing to trade homestead goods or services.

THE BACKYARD HOMESTEADING FOR BEGINNERS GUIDE | 137

Advantages	Disadvantages
• Anyone can trade as long as they have something to trade. • Both parties have power in the trade. • Helps build communication and community. • Less wastage occurs. • No money is exchanged, though, in the cases of high-value trades, there may be some taxes involved.	• Requires a lot of effort and time to find someone to trade with. • You need to find someone willing to trade what you want for what you have. • You'll need to brush up on your communication skills.

There are a variety of things you can trade—especially if you're someone who regularly needs to go into town to pick up items.

• Miscellaneous items: First aid supplies, alcohol, cigarettes, bullets, toilet paper, sugar or honey, batteries, water filters, seeds, lumber, etc.

• Food: Eggs, milk, flour, rice, preserved goods, and animals.

  • In the case of trading animals, ensure that you know your local laws.

• Skills: There are many skills that other people may find handy.

  • Professional skills such as teaching (education or other skills), plumbing, or writing.
  • Mechanical skills can be used to repair or help someone set up a system such as alternative energy.

- Traditional skills, such as candle making, apothecary goods, and sewing, are perfect for bringing more trade to your homestead.

• Services: Cutting down trees, studding (if you have male animals), gardening, cleaning, etc.

To trade fairly, you need to know the value of what you're trading and getting. Do negotiate if you think you can get a better deal. However, always be friendly and polite when bartering, as this will help build your reputation within the community.

COMMON CHALLENGES OF HOMESTEADING AND OVERCOMING THEM

Earlier, the challenges of homesteading were mentioned, and though they seem a little dark, there are many ways you can overcome some of the challenges you'll be facing starting a homestead, as well as for years afterward.

Challenges	Overcoming challenges
Animals	• When in a rural area, getting feed for your animals can be a challenge. However, if you and several homesteaders in the area get together, the shipping costs will be reduced.
Breaking laws unintentionally	• While you may want to dive into homesteading immediately, you mustn't break any laws. • Check all state and local laws before you do anything. ○ If unsure about what animals are allowed in an urban setting, you can check with your local SPCA. • Broken laws can result in buildings being knocked down, animals being confiscated, and possible fines.
Energy needs	• Before leaving the grid, you need to know your energy requirements. • Once you know those requirements, consider the alternative energy resources you'll use. • These systems will have prerequisites (running water, enough space for solar panels, etc.), and when not met, you can't use them.

Finances	• Homesteading is expensive. The more self-sufficient you want to be, the more money you'll have to spend.
	• Be prepared to do thorough research of what you want and budget accordingly.
	• Learn to be creative to make money on your homestead to deal with its running costs.
	• Do your homestead in steps.
	  ○ Change your lifestyle and mindset to one that focuses on self-sufficiency.
	  ○ Start with an urban homestead, and plan for a larger, more rural homestead.
	  ○ Draw up a budget and review it with the projected costs of your plans.
	  ○ Assign priority to the steps you want to take. Getting soil fertile should happen before planting, so don't buy seeds before the soil is ready for them.

Flexibility	• Things will not always go your way, and you may find that if you're too focused on specific goals, your homestead may fail.
	• Learn to be more flexible and have contingency plans for everything you think could go wrong.
	   o The more prepared you are, the better you'll react when something goes wrong.
	   o If it's raining and you can't harvest, go ahead and do an indoor activity such as preserving food.

Health	• Accidents happen, so know where the closest doctor or hospital is. • Learn first aid or take a first responder's course to prepare yourself for stabilizing any situation.
House	• If in a situation where you need to build a home, ensure that it's big enough for what you need but not so big that it can't be powered or heated adequately. • Consult your plans and design a home around what you need, not what you want.
Isolation	• Being lonely will kill your spirit quickly, so reach out to like-minded people and get to know your neighbors. • This is also a great opportunity to have people you can rely on for help and advice. o Knowing your neighbors within walking distance is a must, as these are the people who are close enough to help in an emergency.
Lack of knowledge	• Homesteading can't be achieved with a single book. It takes a lot of research and planning before you acquire the knowledge to develop a working homestead. • Never stop learning and keep acquiring new skills you can use on your homestead or sell.
Location	• If you're unhappy with the property you want to buy for your homestead, don't buy it! o Land is too expensive to have second thoughts about it.

	<ul><li>This is the land that will need to support you, and if you cannot add septic tanks, wells or have access to water, or even have vehicle access, it is useless as a homestead.</li><li>Look at the plans you have to decide if the property you want to buy is worth it.</li></ul>
Neighbors	<ul><li>In the case of urban homesteading, keeping your neighbors happy is key.<ul><li>Smelly compost or noisy chickens will result in complaints.</li><li>Work with your neighbor to determine what is best for both of you.</li></ul></li></ul>
No skills	<ul><li>This goes hand-in-hand with flexibility.</li><li>Learning new skills is vital to start and maintain a homestead.</li><li>Keep learning to make yourself more self-sufficient and live more comfortably.</li><li>Learning to be flexible in itself is a skill.</li></ul>
Not have backup systems	<ul><li>Everything that functions will eventually break; you must have backups or parts on hand to ensure your downtime is limited.</li></ul>
Over-ambitiousness	<ul><li>As eager as you may be, be careful. No one wants to harvest 50 heads of lettuce at once because there is no way to preserve it all.<ul><li>Grow plants in batches to ensure you're harvesting manageable crops throughout the growing season.</li></ul></li><li>The same for animals, you may be keen to have 20 chickens, but can you use all the eggs they lay? Can you afford the large coop, run, and feed needs?</li><li>It's best to start small and grow over time.</li></ul>
Time management	<ul><li>It's going to feel as if you never have enough time to do anything.</li><li>To ensure you get to everything that needs to be done in a year, create a list and break it down into daily, weekly, monthly, and yearly chores.<ul><li>Allow this list to be flexible, but don't procrastinate.</li><li>Developing a routine is a must.</li></ul></li></ul>

Don't allow a challenge to overpower you. As long as you're aware of what lies ahead, you can prepare for it.

EMERGENCY PREPARATION

Emergencies can and will happen; it's only a matter of time. How badly you're affected by them is dependent on how prepared you are for them. It may be easier to deal with problems on an urban homestead, but that doesn't mean it's impossible on a rural homestead.

There are two types of disasters. The first are natural disasters that threaten you and your animals' shelters. There is very little you can do about natural disasters except know how to react when one occurs in an area. The second type is non-shelter-threatening disasters such as injuries, medical issues, and even animal attacks.

Preparing Animals for Natural Disasters

Depending on the natural disaster, you may have to remain where you are or evacuate. This can be a difficult decision to make, not only because you can lose everything, but you also need to consider your animals.

Firstly, think of how the layout of your homestead can offer some protection to your animals—especially when their shelters are built on higher levels, as this will prevent flood waters from getting to them. However, penned-up animals cannot escape if the water or fire reaches them. It may be in the best interest of these animals to release them if you can't take them with you. Many animals still have instincts that will help them survive natural disasters and find a place to weather in. If told to shelter in place, you can have small livestock in the house with you.

However, if your only recourse is to release your livestock if you have to evacuate, mark them to show that they are yours. This can include painting large livestock with your information (waterproof paint), tattoos, or even rings on the legs of chickens. This way, if the animals are picked up later, the rescuers will know who they belong to and return them to you.

Preparing Yourself and Homestead

Even if you don't have to face natural disasters, you may still find yourself in an emergency where you are isolated on your homestead for an extended period. Firstly, you need a way to communicate with the outside world. While many people want to rely on cell phones or satellite phones, there is always a chance that these services can be interrupted. Always have a backup but a radio that you know how to operate and who to contact. Emergency numbers (including the walking distance closest neighbor) should be kept close to the communication devices.

Your home should be stocked with food and water (wells and rivers can become contaminated with floods) or a way to clean water for a minimum of 3–5 days, some homesteaders have enough to last upwards of 30 days. However, this isn't all you'll need. It's a good idea to have emergency fuel (propane or wood), medication, a first aid kit, and essentials such as torches, a crank radio, batteries (with chargers and backup chargers), kerosene lamps, and even a portable propane stove.

When sheltering in place, you need to ensure there's enough food to sustain your animals. You can plan for this by knowing what sort of emergencies and disasters occur in the region you're in.

All these items should also be added to bug-out bags. These bags are only used in a situation where you have to leave your homestead in a hurry. They will need to be serviced once a year, to ensure everything within it is still functional. When forced to evacuate, ensure everyone knows where they need to rendezvous, in case they get separated. Emergency evacuations need to be practiced with children and pets to ensure they know what to do and where they need to go, even if it's to a neighbor.

Emergencies don't need to be scary. Take the time to prepare and practice, ensuring that you're considering your neighbors and family in your plans.

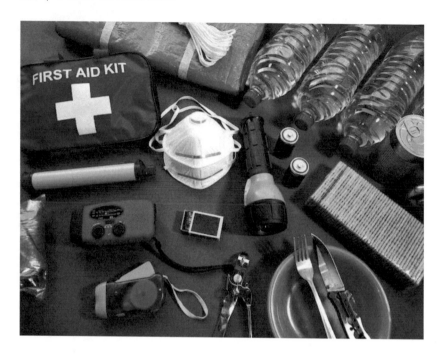

LIVING A SELF-SUFFICIENT AND RESILIENT LIFESTYLE

Once you're ready to make the move to a homestead full-time, away from the hustle and bustle of the urban lifestyle, this isn't the time to get lax in living resiliently. Your homestead needs to be maintained, and there is a lot of work to be done. Here are a few handy tips to keep you self-sufficient:

• Be creative with making money or trading.

• Build structures to last.

• Conserve all resources and reduce waste produced.

- Develop a support system of like-minded people willing to help.

  - These are people who can also teach you new skills.

- If you haven't made the move to an alternative energy source, give it a try.
- Keep learning new skills.
- Learn to be flexible, and if something goes wrong, don't panic.
- Learn to live with less and avoid debt.
- Make all food from scratch.
- Never stop reading.

  - Collect books on a variety of skills you want to learn.
  - Second-hand bookstores are a great place to find helpful guides.

- Practice what you learn, or your skills will diminish.

Remember; this isn't a race. Take your time to get to this point. Never rush major changes in your life. The point is to relieve stress, not make it worse!

CONCLUSION

Starting a homestead isn't about buying a piece of land to go live in the middle of nowhere. It's mostly about planning, budgeting, and mentally preparing to become self-sufficient. Regardless of whether you plan a big homestead or a little one, there are many benefits to be reaped.

Not everyone wants to live a fully self-sufficient lifestyle and often will make smaller changes in their lives to lower their carbon footprint. Urban homesteaders decide how self-sufficient they want to be, as they can decide what food they want to grow and what grids they want to leave.

Homesteading is no longer as difficult as it once was for the pioneers, but that doesn't mean that it isn't a lot of hard work. Homesteads need to be planned and designed before you even start looking for property to own. One of the biggest drawbacks of homesteading is the number of laws you'll have to familiarize yourself with. While you think knowing the state laws are good enough, it's not. Local laws will trump state laws, and if you plan to be an urban homesteader, HOAs can trump even local laws!

If you're not used to being self-sufficient, you'll have to make some changes to the way you live to help reduce waste and have goals of what you want from your homesteading lifestyle. Once you have those goals, you can move on to pursuing them.

A great way to start homesteading is to get a functional garden producing different vegetables and fruit for you. There are many ways to create your garden, from being in the ground to using containers or raised beds. If you are short of space, don't worry. There are many ways to garden vertically, saving you space but not costing you in crops.

You may find that you want a few livestock animals on your homestead. What you get will be determined by the size of the property. Chickens and rabbits make excellent starter animals, and they are easy to look after and don't require a lot of space. With larger animals, you'll need to consider larger pens with pastures and fences to keep them contained.

Once you start producing different crops and animal products, you may quickly find that you cannot eat everything in one sitting. Preservation of food to prevent wastage is a vital skill to learn for homesteading, and it isn't difficult to do. While some preservation methods do require specific pieces of equipment, this isn't true for everything. Whether canning, drying, or curing, there are necessary steps that need to be taken to ensure the preserved food remains safe and edible.

Except for buying land and building a home, the next highest expense will be getting off the grid. Alternative energy sources are a must to power a rural homestead. Which alternative energy you decide to use will depend on your property.

Depending on what state you're in, it's possible to harvest rainwater to use in your garden, or your home with the correct filters and cleaning methods. You can even resort to using gray water for your garden, as long as you're using biodegradable soaps that aren't too harsh on the environment.

Living on a homestead does have many challenges, but with enough perseverance, they can be overcome, and all the rewards can be enjoyed. A homestead can even help bring in some money or other items or services you don't have. Learning how to barter and trade is a great skill to utilize to get what your homestead or you may need.

While homesteading is a simpler way to live, it isn't easier, but you won't know if you're ready to move until you start changing your perception of what self-sufficiency is.

Nothing is stopping you from starting small and learning everything needed before trying to develop a larger homestead. So, why wait? Get some heirloom seeds and get a garden started while teaching yourself how to preserve what you'll grow.

REFERENCES

Accetta-Scott, A. (2019, January 30). *Setting homestead goals | The first year*. A Farm Girl in the Making. https://afarmgirlinthemaking.com/setting-homestead-goals-first-year/

Adamant, A. (2022, May 6). *50+ fruit canning recipes from A to Z*. Practical Self Reliance. https://practicalselfreliance.com/canning-fruit/

Albrecht, J. (2020, August 9). *Jams, jellies and preserves*. UNL Food. https://food.unl.edu/jams-jellies-and-preserves

All Homesteading. (2023, January 28). *How to prepare the raised beds for planting: No dig foundations*. https://allhomesteading.co.uk/2023/01/28/preparing-the-garden-beds-before-planting-step-1-w-video/

Aloi, P. (2023, May 30). *31 raised garden bed design ideas*. The Spruce. https://www.thespruce.com/raised-bed-garden-ideas-4172154

American Veterinary Medical Association. (n.d.). *AVMA animal welfare principles*. https://www.avma.org/resources-tools/avma-policies/avma-animal-welfare-principles

Andrychowicz, A. (2018, February 8). *How to test seed germination with a simple viability test*. Get Busy Gardening. https://getbusygardening.com/how-to-test-viability-of-old-seeds/

Animal Welfare Institute. (2019). *Cattle*. https://awionline.org/content/cattle

Anne of All Trades. (2021, August 27). *15 must have tools for the homestead and garden*. https://www.anneofalltrades.com/blog/15-best-homesteading-tools

Arts Nursery. (n.d.). *Most common plant diseases and solutions*. https://www.artsnursery.com/page/plant-diseases-solutions

Atkins, G. (2023, March 3). *So, how much land do you need to homestead?* The Homesteading Hippy. https://thehomesteadinghippy.com/how-much-land-is-needed-to-homestead/

Avery, J. (2011, October 29). *Integrated pest management*. Farm Homestead. https://farmhomestead.com/integrated-pest-management/

Barnyard Bees. (2017, November 20). *How to keep your honeybees alive through winter* [Video]. YouTube. https://www.youtube.com/watch?v=hRMSBryyUQk

Barth, B. (2018, August 16). *Backyard livestock 101: Chickens, turkeys, goats, and rabbits.* Modern Farmer. https://modernfarmer.com/2018/08/backyard-livestock-101-chickens-turkeys-goats-and-rabbits/

Beatrice, R. V. (2017, March 15). *Homesteading your home: Pro's and con's.* Irrevocable Trust Ultra Trust. https://irrevocable-trust.ultratrust.com/homesteading-your-home-pros-cons.html?amp

Beekeeping Made Simple. (2020, October 27). *Why every homestead should have bees.* https://www.beekeepingmadesimple.com/blog/homestead-beekeeping

Bernauer, A. (2014, May 28). *The financial challenges of homesteading.* Montana Homesteader. https://montanahomesteader.com/financial-challenges-homesteading/

Boeckmann, C. (2023a, May 6). *10 tips for preparing your garden for winter.* The Old Farmer's Almanac. https://www.almanac.com/10-tips-preparing-your-garden-winter

Boeckmann, C. (2023b, June 9). *How to grow vertically in your garden.* The Old Farmer's Almanac. https://www.almanac.com/how-grow-vertically-your-garden

Boye, K. (2016, March 17). *How many acres do you need to board horses?* Rethink Rural. https://rethinkrural.raydientplaces.com/blog/how-many-acres-do-you-need-to-board-horses

Brandon. (2023, March 8). *Garden till methods for a healthy garden.* Jersey Milk Cow. https://jerseymilkcow.com/garden-till-methods-for-a-healthy-garden/

Burgess, R. (2019, March 15). *Living off the grid: 7 challenges and how to overcome them.* A Modern Homestead. https://www.amodernhomestead.com/living-off-the-grid-challenges/

Burke, N. (2022, October 4). *How to build a raised garden bed for just $100.* Gardenary. https://www.gardenary.com/blog/how-to-build-a-raised-garden-bed-for-just-100

Burns, S. (2014, November 7). *Winterizing the farm, garden, or homestead-With a free printable checklist.* Runamuk Acres Conservation Farm. https://runamukacres.com/winterizing-the-farm-with-free-printable-checklist/

Carlson, R. E. (2022, October 28). *Easy guide to growing mushrooms at home.*

Homesteading. https://homesteading.com/growing-mushrooms-at-home/

Carlson, R. E. (2023, January 6). *A homesteader's guide to bartering.* Homesteading. https://homesteading.com/homesteaders-guide-bartering/

Carlson, R. E. (n.d.). *Food preservation methods | Which one is right for you?* Homesteading. https://homesteading.com/food-preservation-methods/

Carpenter, D. (n.d.). *The 5 best alternative energy solutions for homesteaders.* Homestead Launch. https://homesteadlaunch.com/alternative-energy/

Casteleijn, M. (n.d.). *How do you stimulate plant resistance?* Royal Brinkman. https://royalbrinkman.com/knowledge-center/crop-protection-disinfection/how-do-you-stimulate-plant-resilience

composthq. (2021, January 11). *What is good for using as a compost activator?* https://composthq.com/composting/what-is-good-for-using-as-a-compost-activator/

Cooke, W. (2020). *Vermicomposting - nutrient cycling for homesteads (and apartments!).* Permies.com. https://permies.com/t/138252/composting/Vermicomposting-Nutrient-Cycling-Homesteads-Apartments

Coosemans, S. (2021, January 12). *What is homesteading, and how does it relate to permaculture or farming?* Sunny Simple Living. https://sunnysimpleliving.com/what-is-homesteading/

Crank, R. (2019, March 13). *Determining the best renewable energy source for your homestead.* Countryside. https://www.iamcountryside.com/homesteading/best-renewable-energy-source-for-your-homestead/

Culver, B. (2023, May 24). *What is Homesteading? What to know to get started.* An Off Grid Life. https://www.anoffgridlife.com/what-is-homesteading/

Damerow, G. (2022, April 8). *5 homestead fencing mistakes to avoid.* Countryside. https://www.iamcountryside.com/fences-sheds-barns/homestead-fencing-mistakes-avoid/

de Meijere, M. (n.d.). *Becoming more self-sufficient: 9 key strategies to increase self-reliance.* Roots Reconnected. https://www.rootsreconnected.com/becoming-more-self-sufficient

Deanna. (2019a, April 1). *Vermicomposting 101: How to make & maintain a simple worm bin.* Homestead and Chill. https://homesteadandchill.com/vermicomposting-101-worm-bin/

Deanna. (2019a, May 3). *Organic pest control, part 1: How to prevent pests in the garden.* Homestead and Chill. https://homesteadandchill.com/organic-pest-control-prevention/

Deanna. (2020, February 20). *Composting 101: What, why & how to compost at home.* Homestead and Chill. https://homesteadandchill.com/how-to-compost-101/

Deanna. (2021, April 8). *Homemade fertilizer with aloe vera: Soil drench or foliar spray.* Homestead and Chill. https://homesteadandchill.com/homemade-aloe-vera-fertilizer/

Decoteau, M. B. (2020, October 21). *Bartering 101: How to trade for things you need.* Insteading. https://insteading.com/blog/bartering-101/

Deen, B. (n.d.). *Systems of planting.* Agrihortieducation.com. https://www.agrihortieducation.com/2016/09/systems-of-planting.html?m=1

Devon. (2016, February 6). *Creating a medicinal herb garden: Growing herbs for health & wellness.* Nitty Gritty Life. https://nittygrittylife.com/creating-a-medicinal-herb-garden/

Dixon, S. (2021, September 1). *The unsung benefits of homesteading (there's a perfectly good reason for the egg in my pocket.).* Homestead.org. https://www.homestead.org/lifestyle/unsung-benefits-of-homesteading/

Econation. (2023, January 12). *The self-sufficiency mindset.* https://econation.one/blog/the-self-sufficiency-mindset/

Ecrotek. (n.d.). *The basics and benefits of homesteading.* https://www.ecrotek.com.au/blogs/articles/the-basics-and-benefits-of-home-steading

Edie Newsroom. (2009, June 10). *Rainwater or greywater?* Edie. https://www.edie.net/rainwater-or-greywater/?amp=true

The Editors. (2018, July 13). *How to build the ultimate outdoor compost bin.* Good Housekeeping. https://www.goodhousekeeping.com/home/gardening/a20706669/how-to-build-compost-bin/

The Editors. (2023a, February 8). *How to build a raised garden bed.* The Old Farmer's Almanac. https://www.almanac.com/content/how-build-raised-garden-bed

The Editors. (2023b, February 23). *Garden plans for homesteads and small farms.* The Old Farmer's Almanac. https://www.almanac.com/garden-plans-homesteads-and-small-farms

Emily. (2017, November 23). *9 tips for planning the perfect homestead layout.* Accidental Hippies. https://www.accidentalhippies.com/perfect-homestead-layout/

Everett, W. (2021, September 16). *Living off-grid: What it's actually like.* Insteading. https://insteading.com/blog/living-off-the-grid/

Everett, W. (2022, March 21). *Getting started with self-sufficient living (and why it is possible).* Insteading. https://insteading.com/blog/self-sufficient-living/

Extension University of Missouri. (n.d.). *How to build a compost bin.* https://extension.missouri.edu/publications/g6957

Family Food Garden. (2018, December 11). *Design your homestead & backyard farm plans.* https://www.familyfoodgarden.com/homestead-backyard-farm/

Farm Homestead. (2006a, May 13). *Composting.* https://farmhomestead.com/soil/composting/

Farm Homestead. (2006b, May 13). *Vermicomposting.* https://farmhomestead.com/soil/vermicomposting/

Farrow Family Farmstead. (2022, January 17). *10 common homesteading myths.* https://farmsteadfavorites.com/2022/01/17/10-common-homesteading-myths/

Fenley, K. (2022, July 14). *How to make vinegar from scratch.* Cultured Guru. https://cultured.guru/blog/how-to-make-vinegar-from-scratch

Fewell, A. K. (2018, April 19). *Preparing for emergencies on the homestead.* Amy K Fewell. https://thefewellhomestead.com/emergencies-homestead-prepping/

FindLaw. (2016, June 20). *State homestead laws.* https://www.findlaw.com/state/property-and-real-estate-laws/homestead.html

Fisher, S. (2022, July 25). *15 DIY compost bin plans.* The Spruce. https://www.thespruce.com/compost-bin-plans-4769337

Flores, J. (2021, September 10). *Preserving the harvest: Food preservation techniques.* Homestead.org. https://www.homestead.org/food/food-preservation-techniques/

Food Preserving. (n.d.). *Food preserving methods.* http://www.foodpreserving.org/2013/01/food-preserving-methods.html

Food Safety Helpline. (2015, April 8). *What are the different methods of food preservation?* https://foodsafetyhelpline.com/what-are-the-different-methods-of-food-preservation/

From Scratch Farmstead. (2022, May 22). *Setting up A homestead budget for one small income.* https://fromscratchfarmstead.com/setting-up-a-homestead-budget/

Garman, J. (2019, February 1). *How to start beekeeping in your backyard.* Backyard Beekeeping. https://backyardbeekeeping.iamcountryside.com/beekeeping-101/how-to-start-a-honey-bee-farm/

Garman, J. (2021, June 30). *Sheep care on small farms and homesteads.* Timber Creek Farm. https://www.timbercreekfarmer.com/raising-sheep-without-grazing-pastures/

george. (n.d.). *What states is homesteading legal.* Power & Beauty. https://power-beauty.com/what-states-is-homesteading-legal/

Goodrich, V. (n.d.). *Is pickling the same as fermenting?* WebstaurantStore. https://www.webstaurantstore.com/blog/3658/pickling-vs-fermenting.html

Grandin, T. (2006, October 2). *Safe handling of large animals (cattle and horses).* The Beef Site. https://www.thebeefsite.com/articles/706/safe-handling-of-large-animals-cattle-and-horses

Grant, A. (2023, March 8). *Garden layout plans – Tips on layout options for the garden.* Gardening Know How. https://www.gardeningknowhow.com/edible/vegetables/vgen/layout-options-for-gardens.htm

Green Tourism. (2012). *Rainwater and grey water recycling.* http://www.greentourism.eu/en/BestPractice/Details/28

Greens, S. (2022, April 26). *What is vertical garden.* Urban Plants. https://urbanplants.co.in/blogs/news/what-is-vertical-garden

Greg. (2021, January 31). *Homestead size – How big a homestead you need – And why.* Homestead Crowd. https://homesteadcrowd.com/homestead-size-how-big-a-homestead-you-need/

Handyman Homestead. (n.d.). *Small scale solar to power farm needs, rain water collection and distribution is cost effective.* https://handymanhomestead.com/solar-and-water

Harbour, S. (2022, February 12). *Off grid homesteading: 5 challenges and the solutions we're trying.* An Off Grid Life. https://www.anoffgridlife.com/off-grid-homesteading/

Harbour, S. (2023, March 18). *Homestead budgeting: How to get started.* An Off Grid Life. https://www.anoffgridlife.com/homestead-budgeting/

Harris, S. L. (2014, July 12). *Preserving summer: Everything you need to know about canning, drying, and freezing.* Stacy Lyn Harris. https://stacylynharris.com/preserving-everything-you-need-to-know-about-canning-drying-freezing/

HG&H Staff. (2022, September 8). *Off grid solar and wind power for your homestead.* Home, Garden and Homestead. https://homegardenandhomestead. com/off-grid-solar-and-wind-power-for-your-homestead/

Hippies and Heathens. (2023, March 7). *Exploring the advantages and challenges of off-grid living.* https://hippiesandheathens.blog/2023/03/07/exploring-the-advantages-and-challenges-of-off-grid-living/

Homegrown Handgathered. (n.d.). *Homegrown handgathered.* YouTube. https:// www.youtube.com/@HomegrownHandgathered/featured

Homestead House Plans. (2020, September 4). *Real homes for real life.* https:// homesteadhouseplans.com/

Homestead Lady. (2019, September 14). *How to plan and plant a medicinal garden.* https://homesteadlady.com/how-to-plan-and-plant-a-medicinal-herb-garden/

Homestead Lady. (2021, July 12). *Seed saving for the easily confused.* https:// homesteadlady.com/seed-saving-for-the-easily-confused/

homesteaddreamer. (2014, April 11). *5 challenges all homesteaders face.* Homestead Dreamer. https://www.homesteaddreamer.com/2014/04/11/5-challenges-all-homesteaders-face/

James, B. (2018, February 20). *How much land you need for a homestead.* Perma-Resilience. https://permaresilience.com/permaculture-articles/how-much-land-homestead/

Jamie, A. (2022, October 25). *Is homesteading legal? What it involves in modern times.* Why Farm It. https://whyfarmit.com/is-homesteading-legal/

Jeanroy, A., & Ward, K. (2021, August 25). *Food preservation methods: Canning, freezing, and drying.* Dummies. https://www.dummies.com/article/home-auto-hobbies/food-drink/canning/food-preservation-methods-canning-freezing-and-drying-195272/

Jenna. (2019, June 24). *7 great tips for the raising pigs on the homestead.* The Flip Flop Barnyard. https://www.flipflopbarnyard.com/7-tips-novice-pig-farmer/

Johnson, C. (2019, May 15). *Master gardener: Easy tests for seed viability.* The News-Messenger. https://www.thenews-messenger.com/story/news/2019/05/15/master-gardener-easy-tests-seed-viability/3664294002/

Johnson, J. (2018, July 3). *Choosing the best fruits, nuts & berries to plant on your homestead.* Down to Earth Homesteaders. https://downtoearthhomestead

ers.com/choosing-the-best-fruits-nuts-berries-to-plant-on-your-homestead/

Johnston, C. (2021, February 4). *How to plan your garden layout.* Growfully. https://growfully.com/garden-layout/

Josh, & Carolyn. (2022, July 9). *10 things I wish I knew before I started homesteading.* Homesteading Family. https://homesteadingfamily.com/10-things-i-wish-i-knew-before-i-started-homesteading/

Julie. (2019, October 29). *How to choose the best homestead animals.* Simple Living Country Gal. https://simplelivingcountrygal.com/how-to-choose-best-homestead-animals/

Kathi. (n.d.). *How to plan and plant a homestead orchard.* Oak Hill Homestead. https://www.oakhillhomestead.com/2017/03/how-to-plant-homestead-orchard.html?m=1

Kimberly. (2023, April 24). *How much land do you need for a small homestead?* Backyard Homestead HQ. https://backyardhomesteadhq.com/how-much-land-do-you-need-for-a-small-homestead/

Little Homesteaders. (2021, June 2). *Common issues to expect while homesteading.* https://littlehomesteaders.com/common-issues-to-expect-while-homesteading/?amp

Liz. (2018, December 31). *9 essential herbs for the homestead garden.* The Cape Coop Farm. https://thecapecoop.com/9-essential-herbs-for-the-homestead-garden/

Love Property. (2020, May 12). *Survivalists reveal genius tips for self-sufficient living.* https://www.loveproperty.com/galleries/amp/95567/survivalists-reveal-genius-tips-for-selfsufficient-living

Love Your Landscape. (n.d.). *Getting started with vertical gardening.* https://www.loveyourlandscape.org/expert-advice/shrubs-and-flowers/interiorscaping/vertical-gardens/

MadgeTech Marketing. (2021, November 16). *7 ancient methods of food preservation.* MadgeTech. https://www.madgetech.com/posts/blogs/7-ancient-methods-of-food-preservation/?cn-reloaded=1

Masi, F., Bresciani, R., & Satish, S. (n.d.). *Vertical gardens.* SSWM. https://sswm.info/step-nawatech/module-1-nawatech-basics/appropriate-technologies-0/vertical-gardens#

MasterClass. (2021, July 30). *A guide to home food preservation: How to pickle, can, ferment, dry, and preserve at home.* https://www.masterclass.com/arti

cles/a-guide-to-home-food-preservation-how-to-pickle-can-ferment-dry-and-preserve-at-home

Maxwell, S. (2021, January 20). *Types of fences for the homestead*. Mother Earth News. https://www.motherearthnews.com/homesteading-and-livestock/types-of-fences-zmaz06fmzwar/

Mayntz, M. (n.d.). *Tips for raising pigs in your backyard*. Halls Feed and Seed. https://hallsfeedandseed.com/blog/27332/tips-for-raising-pigs-in-your-backyard

McKenzie, P. (2022, September 28). *How to evaluate land for a small farm or homestead*. NC Cooperative Extension. https://vance.ces.ncsu.edu/2022/09/how-to-evaluate-land-for-a-small-farm-or-homestead/

Megan. (2020, March 18). *How to make an easy raised garden bed for your vegetables*. Creative Vegetable Gardener. https://www.creativevegetablegardener.com/easy-raised-garden-bed/

Mindy. (2018, September 29). *6 tips for easy homestead goal setting (that will help you focus)*. Our Inspired Roots. https://ourinspiredroots.com/homestead-goal-setting/

Mitchell, R. (2018, February 27). *How to homestead: Challenges of going off-grid*. The Tiny Life. https://thetinylife.com/how-to-homestead-challenges-of-going-off-grid/

The Modern Homestead. (n.d.). *Cultivating mushrooms*. https://themodernhomestead.us/grow-it/fungi/cultivating-mushrooms/

Morrison, Z. (2020, March 13). *Ten natural tips for healthy poultry growth*. The Poultry Site. https://www.thepoultrysite.com/articles/ten-natural-tips-for-healthy-poultry-growth

My Boreal Homestead Life. (2022, May 22). *How to make chokecherry syrup*. https://myborealhomesteadlife.com/blogs/jams-and-jellies/posts/6977346/how-to-make-chokecherry-syrup

Mya, K. (2019, October 19). *Beautifying your homestead through year-round maintenance*. Homestead Hustle. https://homestead.motherearthnews.com/beautifying-your-homestead-through-year-round-maintenance/

Nicolaides, P. (n.d.). *The blog for self-sufficiency and homesteading enthusiasts*. Self Sufficient Homesteading. https://www.selfsufficienthomesteading.com/

Norris, M. (2021a, February 24). *Planting berry bushes and fruit trees (how many fruit & berry plants per person)*. Melissa K. Norris. https://melissaknorris.com/podcast/how-many-fruit-berry-plants-per-person/

Norris, M. (2021b, May 28). *Homesteading myths and tips for success.* Melissa K. Norris. https://melissaknorris.com/homesteading-myths-and-tips-for-success/

Norris, M. (2021c, December 15). *Staying prepared on a homestead.* Melissa K. Norris. https://melissaknorris.com/staying-prepared-on-a-homestead/

Oblas, D., Schaeffer, Z., & Valeris, M. (2022, March 14). *Everything you need to know to build a simple raised garden bed.* Good Housekeeping. https://www.goodhousekeeping.com/home/gardening/g20706096/how-to-build-a-simple-raised-bed/

Off Grid Living. (2023, January 7). *Making the switch to an off-grid lifestyle: Challenges and rewards.* https://offgridliving.net/making-the-switch-to-an-off-grid-lifestyle-challenges-and-rewards/

Oregon's Wild Harvest. (n.d.). *Winterize your garden: Simple tips from a farmer.* https://oregonswildharvest.com/blogs/on-the-farm/winterize-your-garden-simple-tips-from-a-farmer

Ozawa, M. (2023, April 5). *How to make and fill a raised garden bed.* Martha Stewart. https://www.marthastewart.com/8225103/how-make-plant-raised-garden-bed

Page, T. (2019, April 12). *Easy to grow herbs for your homestead medicinal herb garden.* Homestead Honey. https://homestead-honey.com/easy-to-grow-herbs-for-your-homestead-medicinal-herb-garden/

PerfectWater. (n.d.). *Rainwater harvesting laws you need to know about (2023).* https://4perfectwater.com/blog/rainwater-harvesting-laws

Pierce, R. (2023, March 29). *The ultimate guide to growing your own mushrooms.* New Life on a Homestead. https://www.newlifeonahomestead.com/how-to-grow-mushrooms/

Plantwalker, M. (2023, June 9). *Medicinal herb gardening for beginners.* Chestnut School of Herbal Medicine. https://chestnutherbs.com/medicinal-herb-gardening-for-beginners/

Poindexter, J. (2016, November 23). *28 farm layout design ideas to inspire your homestead dream.* MorningChores. https://morningchores.com/farm-layout/

Poindexter, J. (2017, September 15). *8 basic disaster preparedness tips for every homesteader.* Morning Chores. https://morningchores.com/disaster-preparedness/

Poindexter, J. (2019, May 1). *13 steps for assessing and planning your homestead*

land. MorningChores. https://morningchores.com/assessing-and-plan
ning-homestead/

Rakes, M. (2021, October 15). *18 easy ways to become more self-sufficient.*
Graceful Little Honey Bee. https://www.gracefullittlehoneybee.com/18-
easy-ways-become-self-sufficient/

Reider, J. (2019, August 21). *You should be dry-aging your meats at home—Here's
how.* Food & Wine. https://www.foodandwine.com/cooking-techniques/
dry-aging-meats-how-supper-club

Road to Reliance. (2023, February 25). *Homestead budgeting: Getting started.*
https://roadtoreliance.com/homestead-budgeting/

Rooted Revival. (2022, March 16). *Our 1 acre homestead layout + faq's.* https://
rootedrevival.com/1-acre-homestead-layout/

The Royal Horticulture Society. (n.d.). *How to make a raised bed.* https://www.
rhs.org.uk/advice/how-to-make-a-raised-bed

Sakawsky, A. (2019, October 5). *How to save seeds: Seed saving for beginners.* The
House & Homestead. https://thehouseandhomestead.com/how-to-save-
seeds/

Sakawsky, A. (2022, March 20). *What does it really mean to be self-reliant?* The
House & Homestead. https://thehouseandhomestead.com/what-is-self-
reliance/

Schipani, S. (2019, April 18). *How to prepare for emergencies when you live off the
grid.* Hello Homestead. https://hellohomestead.com/how-to-prepare-for-
emergencies-when-you-live-off-the-grid/?amp

Seaman, G. (2009, April 22). *Choosing land for homestead living.* Eartheasy.
https://learn.eartheasy.com/articles/choosing-land-for-homestead-
living/

Seaman, G. (2018, July 31). *5 indispensable tools for homesteading.* Eartheasy.
https://learn.eartheasy.com/articles/5-indispensable-tools-for-
homesteading/

Seely, H. (2021, September 13). *How to live a more self-sufficient lifestyle: 8 tips
for anyone.* Tamborasi. https://www.tamborasi.com/self-sufficient-
lifestyle/

Sherman, E. (2020, April 3). *12 pickled, cured, and fermented foods from around
the world.* Matador Network. https://matadornetwork.com/read/pickled-
cured-fermented-foods/

Singh, B. (2021, March 23). *Sowing.* BYJU'S. https://byjus.com/biology/sowing/

The SJ Team. (2023, April 19). *What is a self-sufficient homestead & why you should start one.* Sustainable Jungle. https://www.sustainablejungle.com/sustainable-living/self-sufficient-homestead/

Stark Bro's. (n.d.). *Adding fruit to your homestead.* https://www.starkbros.com/growing-guide/article/adding-fruit-to-your-homestead

Steele, L. (n.d.). *How much space does a homestead need for livestock?* Fresh Eggs Daily. https://www.fresheggsdaily.blog/2018/04/how-much-space-does-homestead-need-for.html?m=1

StoneyCreekFarm. (2023, February 24). *Drip irrigation for backyard homesteads.* https://stoneycreekfarmtennessee.com/drip-irrigation-for-backyard-homesteads/

Tamara. (2023, January 22). *10 benefits of gardening for a healthy lifestyle.* The Reid Homestead. https://thereidhomestead.com/benefits-of-gardening/

Tannehill, E. (2020, April 20). *10 ways homesteading improves your health.* The Tannehill Homestead. https://www.thetannehillhomestead.com/homesteading-improves-mental-health/

Taylor, T. (2017, June 14). *10 ways to easily keep your homestead neat and organized.* Morning Chores. https://morningchores.com/organizing-homestead/

Thomas, & Carolyn. (2022, December 5). *What is homesteading?* Homesteading Family. https://homesteadingfamily.com/what-is-homesteading/

Thomas, N. (2022, February 22). *Different types of gardening and soil preparation for homesteaders.* HubPages. https://discover.hubpages.com/living/Different-Types-of-Gardening-for-Homesteaders

Tracy. (2014, August 11). *Self sufficient living on a homestead in 10 easy steps.* Our Simple Homestead. https://oursimplehomestead.com/self-sufficient-living-homesteading/

Tracy. (2015, December 15). *Beekeeping for beginners.* Our Simple Homestead. https://oursimplehomestead.com/beekeeping-for-beginners/

The Tropical Homestead. (n.d.). *Self-sufficient living : 23 skills that help.* https://thetropicalhomestead.com/self-sufficient-living-23-skills-that-help/

Turner, J. (n.d.). *Dual rainwater & greywater harvesting systems to save water (2023).* MindsetEco. https://mindseteco.co/rainwater-and-greywater-harvesting-systems/

University of Minnesota Extension. (n.d.). *Poultry care and management.* https://extension.umn.edu/poultry/poultry-care-and-management

University of New Hampshire. (2018, April 10). *10 easy steps to prevent common garden diseases* [Fact sheet]. https://extension.unh.edu/resource/10-easy-steps-prevent-common-garden-diseases-fact-sheet

Vanorio, A. (2020, March 22). *Bartering for goods and services.* Fox Run Environmental Education Center. https://www.foxrunenvironmentaleducationcenter.org/alternative-energy-blog/2020/3/22/bartering-for-goods-and-services

Walliser, J. (2017, June 26). *Plant diseases in the garden: How to prevent and control them.* Savvy Gardening. https://savvygardening.com/plant-diseases-in-the-garden-prevent-control/

Warwick, S. (2022, August 23). *How to plan a garden – Expert layout and planting advice.* Homes and Gardens. https://www.homesandgardens.com/advice/how-to-plan-a-garden

Webb, B. (2017, March 23). *What's stopping you? Top 10 homesteading myths.* Rural Mom. https://www.ruralmom.com/2017/03/top-10-homesteading-myths.html

Welles, H. (2018, November 5). *Guarding your homestead against harmful pests.* Homestead Hustle. https://homestead.motherearthnews.com/guarding-homestead-against-harmful-pests/

Wild Bluebell Homestead. (2023, April 6). *How to save money by bartering and trading for goods and services as a homesteader.* https://wildbluebell.ca/bartering-and-trading

Will, M. J. (2022, March 16). *How to check if seeds are viable (home germination test).* Empress of Dirt. https://empressofdirt.net/seed-germination-test/

Winger, J. (2019, January 22). *Become a beekeeper: 8 steps to getting started with honeybees.* The Prairie Homestead. https://www.theprairiehomestead.com/2014/05/get-started-honeybees.html

Winger, J. (2020, June 5). *Making and using compost for your garden.* The Prairie Homestead. https://www.theprairiehomestead.com/2020/06/making-and-using-compost.html

Winger, J. (2021, July 24). *How to set homestead goals you'll actually achieve.* The Prairie Homestead. https://www.theprairiehomestead.com/2019/01/set-homestead-goals.html

Winger, J. (2022, July 19). *My favorite ways to preserve food at home.* The Prairie

Homestead. https://www.theprairiehomestead.com/2020/08/ways-to-preserve-food-at-home.html

Winger, J. (2023, January 5). *How to afford a homestead.* The Prairie Homestead. https://www.theprairiehomestead.com/2022/01/how-to-afford-a-homestead.html

Woolbright, M. (2022, August 15). *Vertical vs raised bed gardening.* Vertical Garden Supply. https://www.verticalgardensupply.com/blog/raised-bed-gardening/

IMAGE REFERENCES

Andreas. (2023, February 21). *A hen in a chicken coop.* Pexels. https://www.pexels.com/photo/a-hen-in-a-chicken-coop-15645657/

Broggoli, A. (2019, May 2). *Man holding beehive.* Pexels. https://www.pexels.com/photo/man-holding-beehive-2260932/

Brown, R. (2020, August 18). *First aid and survival kits.* Pexels. https://www.pexels.com/photo/first-aid-and-surival-kits-5125690/

Caniceus, J. (2020, July 27). *Garden vegetables cultivation.* Pixabay. https://pixabay.com/photos/garden-vegetables-cultivation-5440802/

Congerdesign. (2017, July 20). *White cabbage garden.* Pixabay. https://pixabay.com/photos/white-cabbage-garden-2521700/

cottonbro studio. (2020, July 21). *Man in blue denim vest and blue denim shorts standing on green grass field.* Pexels. https://www.pexels.com/photo/man-in-blue-denim-vest-and-blue-denim-shorts-standing-on-green-grass-field-4918149/

Hondow, B. (2016, August 31). *Pupa, lady beetle pupa.* Pixabay. https://pixabay.com/photos/pupa-lady-beetle-pupa-1629692/

kezia. (2020, September 4). *Dwarf goat, kid, goat, baby goat.* Pixabay. https://pixabay.com/photos/dwarf-goat-kid-goat-baby-goat-5538873/

lumix2004. (2016, December 2). *Apples, orchard, apple.* Pixabay. https://pixabay.com/photos/apples-orchard-apple-trees-1873078/

Maggs, M. (2017, August 2). *Vegetable, produce.* Pixabay. https://pixabay.com/photos/vegetable-produce-fresh-sales-2573149/

Nelan, B. (2014, September 19). *Lamb, sheep.* Pixabay. https://pixabay.com/photos/lamb-sheep-fence-farm-animal-451982/

Riether, F. (2022, March 7). *Tomato plants, gardening.* Pixabay. https://pixabay.com/photos/tomato-plants-gardening-flower-pots-7049277/

RitaE. (2016, August 12). *House, farmhouse.* Pixabay. https://pixabay.com/photos/house-farmhouse-old-house-1586435/

Robbins, T. (2015, December 26). *Jam, jar.* Pixabay. https://pixabay.com/photos/jam-jar-christmas-homemade-1106592/

scooterenglasias. (2017, August 29). *South Africa, biltong.* Pixabay. https://pixabay.com/photos/south-africa-biltong-meat-drying-2691938/

StockSnap. (2015, September 5). *Writing, writer, notes.* Pixabay. https://pixabay.com/photos/writing-writer-notes-pen-notebook-923882/

Wei, W. (2018, December 4). *Six potted plants close up photo.* Pexels. https://www.pexels.com/photo/six-potted-plants-close-up-photo-1660533/

Wing, S. (2021, February 7). *Aquaculture, urban vegetable cultivation.* Pixabay. https://pixabay.com/photos/aquaculture-5990462/

Zimmer, M. A. (2014, August 17). *Compost garden waste bio nature.* Pixabay. https://pixabay.com/photos/compost-garden-waste-bio-nature-419261/

Zimmer, M.A. (2016, April 12). *Energy, echo.* Pixabay. https://pixabay.com/photos/energy-eco-solar-wood-photovoltaic-1322810/

Made in the USA
Columbia, SC
23 June 2024

37433541R00093